™

References for the Rest of Us!®

BESTSELLING BOOK SERIES

Do you find that traditional reference books are overloaded with technical details and advice you'll never use? Do you postpone important life decisions because you just don't want to deal with them? Then our *For Dummies®* business and general reference book series is for you.

For Dummies business and general reference books are written for those frustrated and hard-working souls who know they aren't dumb, but find that the myriad of personal and business issues and the accompanying horror stories make them feel helpless. *For Dummies* books use a lighthearted approach, a down-to-earth style, and even cartoons and humorous icons to dispel fears and build confidence. Lighthearted but not lightweight, these books are perfect survival guides to solve your everyday personal and business problems.

> "More than a publishing phenomenon, 'Dummies' is a sign of the times."
>
> — The New York Times

> "...you won't go wrong buying them."
>
> — Walter Mossberg, Wall Street Journal, on For Dummies books

> "A world of detailed and authoritative information is packed into them..."
>
> — U.S. News and World Report

Already, millions of satisfied readers agree. They have made For Dummies the #1 introductory level computer book series and a best-selling business book series. They have written asking for more. So, if you're looking for the best and easiest way to learn about business and other general reference topics, look to *For Dummies* to give you a helping hand.

Hungry Minds™

1/01

Weather For Dummies®

Cheat Sheet

Key Weather Words

Atmosphere: The envelope of gases that compose the air surrounding Earth.

Chaos: A state of a system in which disturbances large and small grow and decay. (The atmosphere is *chaotic,* and so is unpredictable beyond a few days.)

Climate: The average, long-term weather of a place.

Coriolis Effect: The "bending" effect of the Earth's rotation on the path of things in motion in the atmosphere and the ocean. The bending or deflection of its course is to the right in the Northern Hemisphere and to the left in the Southern Hemisphere.

Dewpoint: The temperature to which air must be cooled in order for it to become saturated with water vapor.

El Niño: The tropical Pacific Ocean becomes warmer, and air pressure changes, reducing the strength of east to west winds. These changes can affect weather in many parts of the world.

Equinox: Latin for "equal nights." The time in spring and autumn when the Sun shines directly over the Equator and hours of daylight and darkness are equal everywhere.

Global warming: The idea that the continual buildup of greenhouse gases in the atmosphere is leading to warming of temperatures that could alter climate patterns and seriously disrupt societies.

High pressure system: An area where more air has been added overhead than in surrounding areas. That accounts for higher barometric pressure. Typically, the air enters at high altitudes, sinks, and exits at ground level. The sinking motion causes warming and drying, leaving the clear sky often found in high pressure areas.

Low pressure system: An area of rising air usually marked by cloudiness, often referred to as a storm.

Ozone hole: A thinning of the protective ozone layer in the stratosphere, often observed over Antarctica since the late 1970s during the Southern Hemisphere's spring.

Precipitation: Water vapor that condenses in the atmosphere, falling to the surface as rain, snow, or ice.

Pressure: The weight of the air overhead, exerted in all directions on everything air touches. Horizontal differences in pressure cause winds. Vertical differences in air pressure influence cloud formation and storm development.

Relative humidity: The percentage of the air that is saturated with water vapor at the current temperature. A value that changes with temperature. Air that is saturated at 50 degrees — 100 percent relative humidity — falls to about 50 percent relative humidity when its temperature rises to 70 degrees.

Solstice: The point reached on or about June 21 and December 21 when the seasonal track of sunlight over the Earth reaches its northernmost and southernmost progress.

Stratosphere: The layer of much thinner gases in the atmosphere above the troposphere, between 7 miles and 30 miles in height. It includes the ozone layer. It is called the stratosphere because the temperatures are usually stratified and uniform at this level.

Troposphere: The lowest part of the atmosphere, where all of the weather takes place. Its height averages about 7 miles, ranging from about 5 miles at the poles to about 10 miles at the Equator.

Wind chill: The additional cooling effect of wind blowing on bare skin.

Weather For Dummies®

Cheat Sheet

Clouds by Types

Three common main cloud types form in different layers of the atmosphere, and a fourth common cloud type forms vertically.

High layered (above 17,000 feet)

Cirrus: Delicate white strands of ice crystals, often forming "mares tails."

Cirrostratus: A veil of white cloudiness often covering the entire sky, causing "halos" around the moon and Sun and frequently indicating an approaching storm.

Cirrocumulus: Small white patchy patterns like fish scales and often called "mackerel skies."

Middle layered (6,000 to 17,000 feet)

Altostratus: Drab gray clouds of water droplets that obscure the image of the Sun or moon. They can produce rain and snow.

Altocumulus: A darker, larger pattern of patchiness that may produce a shower.

Low layered (below 6,000 feet)

Stratus: Wispy cloud of fog that hangs a few hundred feet above the ground, often bringing drizzle.

Stratocumulus: Dark gray clouds, often covering the entire sky, which usually do not rain. They form rounded wavelike bands that are broken by blue sky.

Nimbostratus: Low, dark, ragged rain clouds that often bring continuous rain or sleet or snow.

Vertical clouds

Cumulus: Large, billowy "cotton balls" of clouds with dark bottoms and bright white tops that can reach 10,000 feet high. May produce brief showers.

Cumulonimbus: Towering thunderheads, dark on the bottom and white anvil-shaped tops that can extend to 50,000 feet. Often produces lightning and heavy precipitation, including hail, and occasionally tornadoes.

Hungry Minds™

For Dummies: Bestselling Book Series for Beginners

Weather

FOR

DUMMIES®

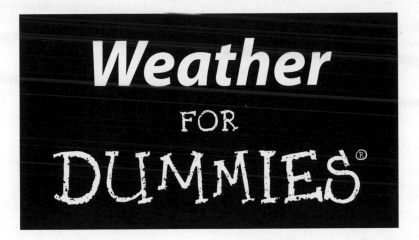

Weather
FOR
DUMMIES®

by John D. Cox

Hungry Minds™

Best-Selling Books • Digital Downloads • e-Books • Answer Networks • e-Newsletters • Branded Web Sites • e-Learning

New York, NY ◆ Cleveland, OH ◆ Indianapolis, IN

Weather For Dummies®

Published by
Hungry Minds, Inc.
909 Third Avenue
New York, NY 10022
www.hungryminds.com
www.dummies.com

Library of Congress Control Number: 00-104218

ISBN: 0-7645-5243-0

Printed in the United States of America

10 9 8 7 6 5

1O/QZ/QY/QQ/IN

Distributed in the United States by Hungry Minds, Inc.

Distributed by CDG Books Canada Inc. for Canada; by Transworld Publishers Limited in the United Kingdom; by IDG Norge Books for Norway; by IDG Sweden Books for Sweden; by IDG Books Australia Publishing Corporation Pty. Ltd. for Australia and New Zealand; by TransQuest Publishers Pte Ltd. for Singapore, Malaysia, Thailand, Indonesia, and Hong Kong; by Gotop Information Inc. for Taiwan; by ICG Muse, Inc. for Japan; by Intersoft for South Africa; by Eyrolles for France; by International Thomson Publishing for Germany, Austria and Switzerland; by Distribuidora Cuspide for Argentina; by LR International for Brazil; by Galileo Libros for Chile; by Ediciones ZETA S.C.R. Ltda. for Peru; by WS Computer Publishing Corporation, Inc., for the Philippines; by Contemporanea de Ediciones for Venezuela; by Express Computer Distributors for the Caribbean and West Indies; by Micronesia Media Distributor, Inc. for Micronesia; by Chips Computadoras S.A. de C.V. for Mexico; by Editorial Norma de Panama S.A. for Panama; by American Bookshops for Finland.

For general information on Hungry Minds' products and services please contact our Customer Care Department within the U.S. at 800-762-2974, outside the U.S. at 317-572-3993 or fax 317-572-4002.

For sales inquiries and reseller information, including discounts, premium and bulk quantity sales, and foreign-language translations, please contact our Customer Care Department at 800-434-3422, fax 317-572-4002, or write to Hungry Minds, Inc., Attn: Customer Care Department, 10475 Crosspoint Boulevard, Indianapolis, IN 46256.

For information on licensing foreign or domestic rights, please contact our Sub-Rights Customer Care Department at 212-884-5000.

For information on using Hungry Minds' products and services in the classroom or for ordering examination copies, please contact our Educational Sales Department at 800-434-2086 or fax 317-572-4005.

Please contact our Public Relations Department at 212-884-5163 for press review copies or 212-884-5000 for author interviews and other publicity information or fax 212-884-5400.

For authorization to photocopy items for corporate, personal, or educational use, please contact Copyright Clearance Center, 222 Rosewood Drive, Danvers, MA 01923, or fax 978-750-4470.

Hungry Minds™ is a trademark of Hungry Minds, Inc

About the Author

John D. Cox is a veteran science writer who has focused his work on weather and climate research. He has written extensively about such climate events as El Niño and such hotly debated climate issues as global warming.

A member of the National Association of Science Writers, he was a 1995–96 Knight Science Journalism Fellow at the Massachusetts Institute of Technology, a world famous meteorological research center. In 1994, he was an invited participant in the research workshop, "The Potential Use and Misuse of El Niño Information in North America," in Boulder, Colorado, under the auspices of the National Center for Atmospheric Research.

John is currently writing *The Weathermen*, a popular history of meteorology, for John Wiley & Sons, Inc.

He has worked for many years as a daily journalist in several bureaus in the western U.S. for United Press International, in London as an editor for Reuters Limited, and in California as a senior writer for *The Sacramento Bee*.

He lives in northern California.

Dedication

To Jan, with love.

Acknowledgments

Thanks first to the many meteorologists and climate scientists at the National Weather Service and elsewhere for their patient guidance through the intricacies of their fascinating science.

Thanks to the leadership of Victor K. McElheny, who as director of the Knight Science Journalism Program at the Massachusetts Institute of Technology inspired a generation of science writers. I am deeply grateful for the personal encouragement of Deborah Blum, a Pulitizer Prize-winning science writer now at the University of Wisconsin, Madison, and of Stephen P. Maran, a multitalented astrophysicist and author of *Astronomy For Dummies*.

To my literary agent, Skip Barker of The Wilson Devereux Company, thanks for the extraordinary support. To artist Scott Flodin, many thanks for the illustrations that grace these pages, and to Jim Reed for some great photographs. And thanks to the editors at Hungry Minds, Inc., especially to project editor Kelly Ewing, and to Karen Young and Stacy Collins.

Publisher's Acknowledgments

We're proud of this book; please send us your comments through our Online Registration Form located at www.dummies.com.

Some of the people who helped bring this book to market include the following:

Acquisitions, Editorial, and Media Development

Project Editor: Kelly Ewing

Associate Acquisitions Editor: Karen S. Young

Editorial Manager: Jennifer Ehrlich

Editorial Administrator: Michelle Hacker

Illustrator: Scott Flodin

Photographers: Jim Reed, UCAR/NCAR, NOAA

Production

Project Coordinator: Amanda Foxworth

Layout and Graphics: Jackie Bennett, Jason Guy, Rashell Smith, Julie Trippetti, Jeremey Unger

Proofreaders: Betty Kish, Marianne Santy

Indexer: Rebecca R. Plunkett

General and Administrative

Hungry Minds, Inc.: John Kilcullen, CEO; Bill Barry, President and COO; John Ball, Executive VP, Operations & Administration; John Harris, CFO

Hungry Minds Consumer Reference Group

    Business: Kathleen A. Welton, Vice President and Publisher; Kevin Thornton, Acquisitions Manager

    Cooking/Gardening: Jennifer Feldman, Associate Vice President and Publisher

    Education/Reference: Diane Graves Steele, Vice President and Publisher

    Lifestyles/Pets: Kathleen Nebenhaus, Vice President and Publisher; Tracy Boggier, Managing Editor

    Travel: Michael Spring, Vice President and Publisher; Suzanne Jannetta, Editorial Director; Brice Gosnell, Publishing Director

Hungry Minds Consumer Editorial Services: Kathleen Nebenhaus, Vice President and Publisher; Kristin A. Cocks, Editorial Director; Cindy Kitchel, Editorial Director

Hungry Minds Consumer Production: Debbie Stailey, Production Director

♦

The publisher would like to give special thanks to Patrick J. McGovern, without whom this book would not have been possible.

♦

Contents at a Glance

Cartoons at a Glance

By Rich Tennant

page 7

page 77

page 165

page 253

page 303

Fax: 978-546-7747
E-mail: richtennant@the5thwave.com
World Wide Web: www.the5thwave.com

Table of Contents

Introduction

∙∙∙

Weather is a big part of life. Certainly it is part of life in the sense that weather is something that everyone experiences more or less directly every day. And certainly weather's extremes of storm and heat are something that most people have to put up with at one time or another.

But weather is part of life in a bigger sense. It is part of life in the same way that the air that you and I breathe is part of it. Often weather gets talked about as something that interferes with my travel plans or interrupts your picnic, but that is not the point. Without weather, *there is no picnic*. No food, no forest, no flowing fresh water.

What's going on up there when the wind blows, when the clouds roll in, when the rain falls and the lightning flashes? To wonder about these things is to share some thoughts with the first people who poked their heads out of a cave and looked up into the dark sky of a violent storm. It is part of being human. This wondering about the weather came long before there was reading and writing and science, and long before there were reasonable explanations for these things. Some of the old explanations, you wouldn't believe. The wind, the clouds, the rain, and the lightning make a lot more sense to the likes of you and me than they used to, but when all is said and done, you have to admit, still they are wonderful.

About This Book

The reasonable weather explanations that separate you and me from the folks poking out of the cave are part of the modern knowledge specialty of *meteorology,* which is the five-dollar word for the science of weather and climate. That's what this book is all about. Weather scientists know the answers now to the basic questions about the changes that take place in the sky and plenty more.

Already you know more than you probably think you do about the weather. Phrases like "low pressure system" and "high pressure ridge" have become familiar, even if not quite understood. And images from space satellites of enormous arms and blotches of cloudiness slowly swirling over the surface of Earth appear on television screens as familiar as the faces of friends. Already you are ahead of people who wondered about the weather some 40 years ago before the satellites went into orbit and made the great size of storms so obvious.

So even before you tackle the details of the comings and goings in the air over your head, some congratulations are in order. In most times past, when people wondered about the weather, they were scared to death. They were frightened by the storms, and when they asked questions about them, they were frightened by the answers they got. If I told you it was the magic of the witch doctor, or the fact that the gods are angry, now you would laugh at me. You and I have come a long way, baby.

There is no right way to read this book, and no wrong way to read it either. You can read it straight through from the first page to the last, but you don't need to. You don't need to read Chapter 1, for example, to get a grip on the subjects covered in Chapter 2. Browse through it or start anywhere you like. If there's something about how weather happens that's been bugging you, just jump in and check it out. *Weather For Dummies* is your ready reference on the subject.

Foolish Assumptions

To write this book, I had to make some assumptions about you. I think you are somebody who enjoys watching the changes that take place in the sky from day to day, or month to month. You take some satisfaction in knowing what's behind these changes. You like to know the meaning of the words you hear on the daily weather reports simply because you like to know the meaning of the words you hear. And from time to time, you have some questions about how the weather works.

You are a consumer of weather information. You are not a mathematician. You are not a weather scientist or forecaster. You have a natural curiosity about the weather, and a healthy respect for it. But you are not crazy-in-love-with-it like a storm chaser who runs out the door with a video camera at the first word of severe thunderstorms nearby. You are not a "weather geek," someone who really wonders about the weather *a lot* and who devours every bit of information that they can find on the subject — although maybe you are a weather geek and you just don't want to admit it yet. If this is the case, your secret is safe with me!

How This Book Is Organized

This book is divided into parts that break the big meal of weather science into easily digested portions. Here's how it goes:

Part I: What in the World Is Weather?

Weather For Dummies begins with weather science's most popular finished product: the daily forecast. Without all the numbers and equations, this part describes what goes into making a forecast and understanding what it means. It lays out the terms that apply and the circumstances that make up weather emergencies.

Why is there weather? What basic forms does it take? This is where you find the answers to these questions. It explains why there are storms. It describes precipitation in all of its shapes and sizes. Here you get the idea of air masses meeting along fronts like opposing armies.

Why are there seasons? In this part, you get the big picture look at what makes the seasons and why they come around the way they do.

Part II: Braving the Elements

Weather is a very popular subject when big storms are brewing or when things like summer temperatures are getting to be extreme. These are the weather celebrities that get all the media attention. And in this part, you find a chapter devoted to hurricanes, perhaps the biggest weather celebrities of all.

But behind every storm and every heat wave and every cold snap is a cast of characters that are responsible for the whole production. They make the winds blow. They form the clouds.

This part takes a look behind the scenes of weather and describes its basic elements. What is air pressure all about, and why is it such a big topic of discussion during the weather forecasts? This part answers that question and more and explains how air pressure drives the winds.

Do you know the names of the clouds? Can you tell one type of cloud from another? Here, you get the lowdown on all forms of clouds. And there are two pages of color photographs devoted to the basic cloud types that are spelled out in this part.

Another question keeps coming up: Who the heck is this El Niño, anyway, and what exactly do he and his sister have to do with the weather? In this part, find out all about El Niño and other climate conditions that make one winter different from another.

Part III: Some Seasonable Explanations

Unless you live in the Tropics, near the Equator, where the Sun is high in the sky all year long, or near the polar regions, where the Sun's rays never get very warm, the different times of year have different weather personalities. The different seasons bring different kinds of storms. And fair weather has a different feel to it from one season to the next.

This part looks at the story of weather the way it presents itself to people like you and me who live in the middle latitudes between the Tropics and the poles. It begins with the big storms of winter and focuses on the tornadoes of spring and the thunderstorms and temperature extremes of summer, and it takes a good look at autumn.

Take a look here at the seasons and see how different they are from one side of the United States to another and see what makes these differences so great.

Part IV: The Special Effects

There's a lot going on in the sky. Unfortunately for people who live in many major cities, the sky over their heads is clogged with extra gases and other material that has been dumped into it. Weather doesn't put that stuff up there, of course, but it sure has a lot to do with how bad it gets. In this part, read all about it.

When the sky is clear of pollution, some marvelous things are going on. Effects like rainbows and sun dogs and haloes that form around the Sun and the moon have been drawing rave reviews as long as people have been looking up. This part describes how the atmosphere bends the light and plays all sorts of tricks on your eyes.

Are you thinking about getting up close and personal with the weather? This part describes cool weather experiments and famous weather experimenters and takes a look at what you need to set up your own weather station.

Part V: The Part of Tens

There is weather, and then there is weather. Once in a while, a storm or climate event like a drought comes along that is so terrible that it is remembered from one generation to another. It makes history. This part takes a look at the storms of the 20th century in the United States and around the world that made the weather Hall of Fame.

Before there was weather science, there were other ways of trying to figure out what the atmosphere was up to. This part gives you a good look at weather lore, some of the famous sayings and proverbs and signs that have been passed down through the ages.

Appendix

Where do you go from here? Weather data and other information about the weather is a huge part of the Internet, and in the Appendix, you can find a list of major weather Web sites to get you started in the right direction.

Icons Used in This Book

In the pages of *Weather For Dummies* are symbols that alert you to certain kinds of information. They help you sort through the wide variety of facts and details and put them in your own order. Here's what these symbols mean.

This icon lets you know about a concept, or big idea, that is not just a detail about the weather, but is a whole train of thought on a subject. Big ideas are not complicated. In fact, they are simple. They're important, or big, and worth checking out, because they help explain a lot of details.

Some words are just weather words. There are a whole lot of special weather words that scientists use all of the time when talking to each other, and this book avoids most of them. The ones you find at this symbol are included because they are helpful or interesting.

A lot can go wrong out there in the weather. Some of it is harmful to your health, and some of it is downright dangerous. That's where bad weather gets its reputation. When you see this symbol, look for some tips about what to do if you are hurt by the weather.

Some kinds of information are valuable because they make complicated things easy or they help cut through a lot of detail to a useful idea. That's the kind of thing this symbol points out, an idea that makes things a little quicker or easier.

A lot of details are useful only to a specific subject, but some things are valuable to keep in mind because they help explain a variety of things. That kind of good-to-remember information is what this symbol identifies.

Some weather situations are so dangerous that they should be avoided always. Most of the dangers are pretty obvious, but not all of them. This symbol alerts you to extreme weather conditions where dangers are clear and present.

Don't be alarmed by this nerdy-looking guy. The technical stuff included in this book is not really the heavy-duty number-crunching kind of thing that weather scientists do once in awhile. This symbol alerts you to stuff that's just a little more technical than the rest.

Where to Go from Here

Go outside. I mean it. You've been spending too much time indoors, anyway, so close this book temporarily, tuck it under your arm, and head out the door. Go outside and give your sky a good looking-over. It's your sky, and your weather, because nobody else sees it or feels it exactly like you do. Do you see clouds up there? Do you know how they form or what their names are? Do you know how much fun it is to start practicing identifying the clouds in your sky? If you don't, it's time to come back inside and open *Weather For Dummies* again. Chapter 5 is a good place to start.

Part I
What in the World Is Weather?

In this part . . .

When human beings first looked up into the sky, chances are they saw the clouds before they saw the stars.

The road to understanding the weather has been as long and as hard as the road to understanding the heavens above it. It is so thin, so close to you, this layer of gases that makes the winds and clouds and storms, and yet still there are mysteries and surprises.

But everyday now, the fruits of this effort of understanding it is laid before you and me in a daily weather forecast that is — amazingly, when you think about it — pretty darned accurate!

In this part, I take a closer-than-usual look at the daily forecast, the work-in-progress of a big and remarkable science. And then I begin the plunge into the details, big and small, that make the weather.

Chapter 1

Forecasts and Forecasting

Accurately forecasting weather is a very hard thing to do. In fact, a 100 percent accurate forecast is impossible. (Find the sidebar "The Butterfly Effect" in this chapter and see why.) The local forecast may be easy and breezy, but there's a lot of hard science and heavy-duty number-crunching behind it. My people at the Go Figure Academy of Sciences (GoFAS) tell me that *meteorology,* the study of the atmosphere, is harder than rocket science. (Check out the note about GoFAS for the real lowdown on this imaginary institution.)

Think of it this way: The first thing that a scientist wants his rocket to do when it blasts off its launch pad is to get out of the *atmosphere,* to leave behind all the turbulent mess of swirling gases surrounding Earth. (As you may have noticed, a rocket that *doesn't* do this is quickly in big trouble!) As soon as it can, it enters the quiet, predictable vacuum of space.

A meteorologist, on the other hand, never gets out of this mess of blowing gases we call the atmosphere. This blowing and shifting atmosphere makes the job of a meteorologist a lot more complicated. A research meteorologist not only has to know everything a rocket scientist knows about the physical laws of motion and mass, about gravity and what-have-you, but also about chemistry and fluid dynamics.

And the mathematics, well — let me put it this way: Meteorologists have bigger computers.

Some of the most powerful supercomputers in the world are devoted to the job of helping to figure out what the weather is going to be like one day to the next. You wouldn't believe everything that goes into making a modern, state-of-the-art, accurate weather forecast. But, in this chapter, I'm going to tell you anyway!

Forecasting Prophets

Where does your weather forecast come from? This is a trick question. Most people get their weather forecasts from television or the radio stations that may employ their own meteorologists or from newspapers that pay forecasting firms that supply a page of forecasts, maps, and other graphics. Private weather forecasting companies and meteorologists employed by the media supply the weather information that is seen or heard by about eight out of ten people in the United States.

But where does *that* information come from? That's the trick part of the question. The answer is that a lot of it comes from Maryland, from the suburbs of Washington, D.C. That is where the National Weather Service supercomputers do the work that is the technical backbone of all the weather forecasts that you read and see and hear. You could think of it this way: The National Weather Service is the wholesaler of basic forecasting information, and the private meteorologists and media weather specialists are the retailers.

Private weather forecasting has become big business, employing about 4,000 meteorologists around the country, according to one estimate, including about 1,300 in the media. Annual sales have been estimated at more than $400 million. The taxpayer-supported National Weather Service is still the biggest employer of meteorologists, with about 2,400 on its payroll, and still provides basic weather forecasts everywhere across the country.

Private companies have found a large and growing market for more specialized forecasts for many companies in a variety of weather-sensitive industries as well as the special needs of different media — newspapers, radio, and television — which may not employ their own forecasters. Big customers for special forecasts include travel and transportation companies such as shipping lines and railroads, as well as highway agencies. They include oil exploration and construction companies and agricultural concerns, and even sports teams who want to know what the weather will like at the stadium at game time.

Making a Forecast

While weather information comes in many voices and forms these days, all meteorologists begin their days with certain things in common, just a keystroke away.

The Go Figure Academy of Sciences

Weather science can be a complicated and difficult subject, so I figured *Weather For Dummies* ought to have its very own think tank. So, well . . . I made one up. It's the Go Figure Academy of Sciences (GoFAS), and it's all mine. It can be yours, too, if you want it. I took the best people I could find and put them to work in my own place.

It looks a little like the World Weather Building that the National Weather Service occupies outside of Washington, D.C. (see figure). It also looks a little like the Massachusetts Institute of Technology, except my dome comes to a sharper point. Anyway, I think it looks pretty great!

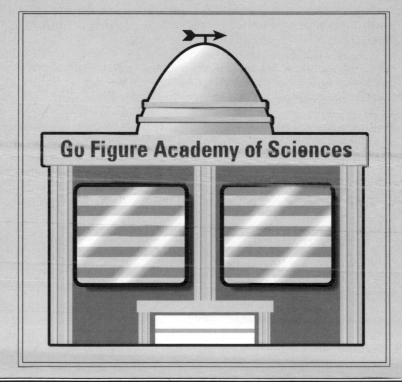

On their computer screens are

- ✔ The latest forecasts issued by the National Weather Service.
- ✔ The latest results of the computer forecasting models supplied by the National Weather Service.
- ✔ The latest images from satellites supplied by the National Weather Service's parent agency, the National Oceanic and Atmospheric Administration.

Behind the scenes, in the World Weather Building in Camp Springs, Maryland, people who work for the National Weather Service play big roles in the daily weather forecast even if you don't hear their voices or see their faces. Television meteorologists compete with one another for your viewing pleasure, with their different styles and presentations and engaging personalities, but they all have the computing and forecasting resources of the National Weather Service on their sides. (And so do you.) It's not a bad way to begin the day.

The basic tool for making modern weather forecasts is the supercomputer. For the outlook beyond the short range of the next several hours, all savvy meteorologists, private and public alike, check out the forecast data supplied by the national and international Numerical Weather Prediction computer models.

The output of these incredibly complicated models are not seen by you and me when the forecast is delivered, but they are a very important part of the forecasting process. But other forecasting techniques still play a role.

- ✔ **Computer modeling.** Day in and day out, the statistics that come out of large and powerful software programs run on supercomputers are the single most important ingredients in the making of the average public weather forecast. This computer process is known as *Numerical Weather Prediction*. At the National Weather Service's World Weather Building near Washington, D.C., data is fed into a software program that acts like a virtual, or model, atmosphere.

- ✔ **Comparing *statistics*.** The big national and regional forecast models don't always pick up the fine details of the landscape such as nearby lakes and small mountain ranges that can affect local weather. An individual local forecaster's knowledge and skills fill in these gaps.

- ✔ ***Observing* current conditions.** Pictures of local conditions are a big part of the daily weather show on television. Observing local conditions is not as important to local weather forecasting as it was before computer forecasting models got so good. Still, it is an important part of any forecasting process. When local weather becomes hazardous, observing current conditions becomes critical. Radar screens, satellite images, and other data are studied. This method is most important for *short-range forecasting* (deciding how weather will change in the next six to 12 hours), and for *nowcasting*, predicting weather for the next one minute to four hours.

What is going to happen in this area during the next 12 hours? (By the way, did you know that in the neighborhood of 90 percent of the members of the American Meteorological Society are male?) The forecaster has to try to answer this question every time he issues a new regularly scheduled public forecast. (Imagine how it must be to have your professional judgment *about the future* put to such a public test so often!)

For special short-range forecasts, such as impending severe thunderstorms, a local forecaster is a busy person. The National Weather Service Storm Prediction Center issues "watches" that alert the public to the possibility of trouble. But then it's up to the local forecaster to keep a close eye on things. On the lookout for damaging hail, local flooding or tornadoes, she relies heavily on local radar observations and uses her own training, experience, and understanding of local conditions to decide whether to issue a public warning. (The section "We interrupt this program . . ." explains the different watches and warnings.)

For forecasts beyond the 12-hour period, and especially for general "outlooks" of weather beyond the next three to five days, the mix of forecasting tools may change. The projections of a variety of computer models still are consulted, but the climatology of a place — its average weather for this time of year — is weighed more heavily. This approach focuses on the question, "What usually happens here this time of year?"

Take what is happening now . . .

The weather forecasting process begins with the need to know what the atmosphere is doing *now*. This information is the data that describes what computer modelers call the "initial conditions" that are the starting points for their powerful forecast software. You might think of it as a nowcast. This information has to be as absolutely accurate and as detailed as possible. Forecast modelers know that the slightest error in their description of the current state of the weather can fairly quickly lead to large errors in the forecast. If the data describing initial conditions is a little wrong, the forecast can be very wrong. (That's what the sidebar "The Butterfly Effect" later in this chapter is all about.)

Weather forecasting is such a complicated and difficult business because it is trying to predict the behavior of a system that has many features that are all changeable, all of the time. Observations of every sort need to be gathered. How warm or how cold is it? Which way is the wind blowing? Is it cloudy, or is it clear? Details about all these features and more are gathered from every source available.

From the ground, these measurements come from human weather observers and from instruments on automatic weather stations. From the air, they come from weather balloons and airplanes and satellites orbiting the planet in space. From the sea, they come from ships and from instruments on anchored moorings and drifting buoys, as well as satellites. (See the section "Tools of the Trade," later in this chapter, for more about these measurement tools.)

What's in a name?

When you think about it, *meteorology* is a funny word for weather science, isn't it? A *meteor* is an object from space, usually a tiny bit of comet dust, that leaves a flash as it burns up in Earth's upper atmosphere. What exactly does that have to do with weather? The answer is exactly nothing!

But Aristotle, the great Greek philosopher, didn't know all of this at the time he first used the word *meteorology* back around 350 B.C. Everything that happened above the earth was considered astronomy in those days, and Aristotle was trying to define a new science. Astronomy was the study of all the stuff that goes on in the distant heavens, he figured, and meteorology was the study of the stuff that happens closer to Earth.

The Greek word *meteoron* means something that falls from the sky. (Hence, the phrase: "Don't be such a meteoron.") Anyway, what Aristotle had in mind, mostly, was weather, the study of rain and snow and hail. But he also threw in such things as comets and earth-quakes. Go figure....

Later, of course, comets were given back to astronomers, and earthquakes became part of the science of geology.

Everyday, 24 hours a day, data from thousands and thousands of weather observations around the world is streaming electronically into National Weather Service computers at the World Weather Building in Camp Springs, Maryland. With these millions and millions of bits of data, the computers are constantly updating and refining their highly detailed descriptions of the current state of the weather.

... And add a little future

Computer models, or software programs, of numerical weather prediction divide the atmosphere over the surface of the Earth and above it into imaginary individual blocks, or *gridpoints.* Different models use different techniques to force changes in their virtual atmospheres. Some are better at depicting the progress of one type of weather change, and some are better at others.

Each of these separate blocks of air has features, such as temperature, pressure, wind, and humidity, with their own values. The computers take the data of all these values from all these blocks of air and apply a set of equations that represent basic physical laws. (No, I know you must be deeply disappointed, but the only equation you are going to find in this book is 1 *Weather For Dummies* = 0 equations.)

Anyway, the output of this incredibly intensive process of computation produces a new value for each key weather feature at each gridpoint. But each complete computation moves the atmosphere forward into the future only a few minutes.

The Butterfly Effect

The atmosphere has a physical characteristic, like a personality trait, that sometimes drives forecasters to distraction, especially as they try to predict what the weather is going to be like beyond a few days.

All disturbances in the atmosphere grow and decay. Some disturbances are big enough to be measured — and "seen" by computer models — and predicted. Other disturbances are too small. In such a *chaotic system,* very small disturbances can lead to big disturbances over time.

When he discovered this characteristic of weather in the early 1960s, meteorologist Edward Lorenz at the Massachusetts Institute of Technology came up with an interesting way to describe the problem. Imagine a butterfly in the jungles of the Amazon fluttering its wings and setting in motion a subtle whirl of breeze that travels and magnifies through the atmosphere over time. Farther and farther it goes, bigger and bigger it grows. Two weeks later, this breeze results in a tornado over Kansas. This is known as *The Butterfly Effect.*

Should The Butterfly Effect be taken literally? No. But forecasters ever since then have taken very seriously the big idea behind it. The atmosphere's way of doing things is on a scale that is always going to be finer, or more exacting, than the scale for the data on initial conditions that goes into the numerical forecast models. This means that, no matter how much computer power is thrown into the job, it is simply not possible to create a 100 percent accurate forecast, or a detailed weather forecast beyond about two weeks.

And so the whole process starts all over again. The computer begins analyzing the data in each block with the new values it has forecast, nudging the weather system forward a few minutes at a time until the desired length of weather forecast is achieved. The experts say that to produce a weather forecast of a few days, modern computer models may need to complete more than *one trillion* calculations (1,000,000,000,000).

The forecast method known as Numerical Weather Prediction has led to some pretty impressive improvements in recent years, but nobody expects its computer models to make individual human forecasters obsolete anytime soon. Even with all that computing power, the numerical prediction models are not detailed enough for every weather forecasting purpose. For example, the models don't very accurately account for the effects on weather of local landscape features such as mountains or lakes. And thunderstorms, which look pretty big when they're coming at you, are too small and too short-lived even to show up on the big numerical models.

And another thing about all this computing power and these state-of-the-art weather forecasting models — did I mention that they are not always accurate?

"We Interrupt This Program . . ."

Besides the regularly scheduled daily forecasts, National Weather Service offices around the country issue special weather statements and watches and warnings and advisories appropriate to local circumstances.

They are issued for such things as tornadoes, severe thunderstorms, floods, and winter storm conditions such as blizzards, heavy snow, ice storms or freezing rain, high winds, and duststorms.

When threats to public safety are imminent, bulletin warnings are issued — only by the National Weather Service — and a special communications network known as the Emergency Alert System is activated. When this happens, local radio and television broadcasters interrupt their regularly scheduled programs to pass them along. Every year, the National Weather Services issues between 45,000 and 50,000 severe weather warnings. Whatever the hazard, the differences between these various levels of public notices can be worth knowing.

- ✔ A *Special Weather Statement* often is issued as a "first alert" to the possibility of significant weather. This kind of "heads up" also is issued when forecasters see the likelihood of such things as thunderstorms with small hail, which may not be life-threatening but could make conditions temporarily hazardous.

- ✔ A *Watch,* such as a tornado or severe thunderstorm watch, is issued when dangerous weather conditions like lightning, large hail, and damaging winds are possible for the next several hours. It's time to be on your toes.

- ✔ An *Advisory,* such as a winter weather advisory or wind advisory, is issued when conditions are not life-threatening but still worth keeping in mind. These are especially valuable to travelers in areas experiencing such hazards as snow or winds or fog. Maybe it's time to slow down and think ahead.

- ✔ A *Warning* is issued when potentially dangerous weather is possible within a matter of minutes and residents should seek shelter. A warning means no fooling around — it's time to take some action.

- ✔ A *Severe Weather Statement* often follows up on a warning, to cancel it or modify the area of concern. Also, this statement might alert residents to the presence of such hazards as funnel clouds that are not expected to touch the ground.

NOAA Weather Radio

Timely weather information is readily available to anyone with an FM or AM radio or television set, and most National Weather Service forecast offices and private forecasting concerns maintain World Wide Web pages on the Internet. But what happens when things go badly? If it's emergency preparation you're thinking of, a battery-powered NOAA Weather Radio belongs in a disaster supplies evacuation kit.

This official weather emergency radio service is the only direct link the National Weather Service has to the public. It operates from about 500 transmitters in 50 states and U.S. territories on seven frequencies in the VHF band, ranging from 162.400 to 162.550 megahertz. These frequencies are outside the normal range of AM and FM broadcasts, although some manufacturers are including NOAA Weather Radio as a special feature on some receivers.

Nearly every National Weather Service office operates at least one NOAA Weather Radio transmitter broadcasting weather information 24 hours a day. The average range of these transmitters is 40 miles.

During severe weather, a tone alert can be activated to cause radios equipped with the alert feature to sound an audible alarm. If you're in the market for a weather emergency radio, check to see whether it has this special tone alarm feature.

Water, Water, Everywhere . . .

The No. 1 weather-related killer in the United States is flooding — not the winds of a hurricane or a tornado — and the especially dangerous circumstance of high water has its own set of National Weather Service bulletins.

A flood is called a *flash flood* because it is sudden. It takes place within a few minutes or a few hours of heavy rainfall or some other event like a dam break or a river levee failure. Most flash flooding is caused by torrential rains from thunderstorms or the rains of hurricanes or tropical storms.

Every state in the United States has been hit by flooding of one kind or another. Rivers flood sometimes in the spring when runoff from heavy rains combines with water from melting snow, although floods can happen any time of year. Along coastlines, the winds of powerful storms can generate big waves and high tides and storm surges that cause coastal flooding. The streets of every city can fill with water of urban flooding when circumstances are just right — or just wrong!

The National Weather Service issues these special flood warnings:

✔ A *Flash Flood Watch* or *Flood Watch* is issued when flooding is possible. Be alert to signs of flooding and be ready to evacuate to higher ground.

✔ A *Flash Flood Warning* or *Flood Warning* means flooding has been reported or is imminent. It is time to act, and to act quickly, to save yourself.

✔ An *Urban and Small Stream Advisory* alerts you to the fact that flooding is occurring on some small streams or streets and low-lying areas such as underpasses and storm drains.

✔ A *Flash Flood Statement* or *Flood Statement* contains follow-up information about a flood event.

Flavors of Forecasts

Depending on where you live and what you do, the weather forecast that wakes you up in the morning probably lets you know what basic conditions to expect of the day. Sometimes it delivers important information about your safety or comfort, and every day millions of people rely on it without giving it much thought.

In the newspaper, on the radio, on television and on the Internet, the basic weather forecast is usually routine, simple, and to the point. It directs itself to the questions that most people want answers to as they start their day. How warm or how cold is it going to be? Will it rain or snow? Is a storm on the way? Will the wind blow? Before this day is over, will I need an umbrella or a light jacket or a heavy coat?

The National Weather Service and other agencies and private companies also prepare daily a number of specialized forecasts. They provide important weather information about the day, or the next few days, to a variety of special audiences around the country. Private forecasting companies supply specially tailored forecasts for hundreds of special users such as utility companies, construction companies, hotels, ski resorts, and motion picture studios.

Here are some of the more common special forecasts.

Agricultural forecasts

A farmer wants to know everything you want to know about tomorrow's weather, but he also has to keep his eye out for a few other special things. Much depends on the average seasonal conditions, or climate, of the region, and, of course, on the kinds of crops in the field.

All farmers are on the lookout for frost, because of the damage it causes to crops. In areas with large citrus orchards and vineyards, where freezing temperatures are rare, growers use large outdoor fans and heating devices to keep their trees and vines from freezing. These fans cause the coldest air, which is nearest the ground, to circulate with warmer air above it.

Another key indicator for many farmers is the combination of sunshine, heat, humidity, and wind conditions that let them know how much evaporation of moisture from their fields to expect. This *evapotranspiration rate* is closely watched by farmers who irrigate their fields.

The U.S. Department of Agriculture and many private companies provide farmers with weather forecasts tailored especially to their local needs around the country.

Aviation forecasts

Forecasters provide specific nowcast guidance for Federal Aviation Administration air traffic controllers to use in advising pilots for flight planning and flight operations. (If you're unfamiliar with nowcasts, see the section "Take what is happening now . . ." earlier in this chapter.)

Also, many of the automatic and manually operated weather stations described in the section "Tools of the Trade," later in this chapter, are located near the touchdown zones of airport runways. The readings from their instruments are automatically broadcast to pilots.

Using communication satellites, the National Weather Service now beams a new radio broadcast system called the World Area Forecast System that is designed for the use of pilots of commercial airliners around the globe.

This tool gives them the best information available about such things as upper-level winds and temperatures, which is crucial to aviators (and their passengers!).

Marine forecasts

No group is more dependent on weather conditions for their safety and well-being than *mariners* — professional sailors, recreational boaters, and commercial and sports fishermen.

Monitoring coastal and offshore conditions, the National Weather Service forecast offices in coastal areas around the United States and in the Great Lakes region issue a variety of specialized forecasts and warnings for professional mariners and recreational boaters.

Marine forecasts let boaters know what kinds of winds to expect, whether the sea will be calm, and, if not, the height and direction of rolling swells. Small craft warnings for strong winds or fog conditions and other hazard advisories are issued in areas along the seacoasts and the shores of large lakes.

River forecasts

National Weather Service forecasters work closely with *hydrologists,* who are water experts. A weather forecaster estimates how much rainfall is likely from a particular storm, and the hydrologist figures out how much that precipitation will cause certain rivers and streams to rise.

Forecasters also keep operators of federal and state dams informed as major storm systems approach large river systems. Sometimes a dam is good for saving water when there is too little, and sometimes a dam is good for saving a downstream city where there is too much. If they know a big storm is on the way, dam operators sometimes release more water from their spillways to make more room in their reservoirs to hold floodwaters.

Regional River Forecast Centers are spread around 13 locations in the United States. These centers are where National Weather Service forecasters prepare river and flood forecasts and warnings for 3,000 communities around the country.

Fire forecasts

A corps of specially trained weather specialists keep their eyes on the skies above national forests and other regions of their nation where wildfires are a major threat. These specialists are experts on the effect that weather has on the risk of a fire starting. They keep an eye on the dryness of the soil and vegetation, on temperatures, and on winds and humidity.

They also have special knowledge about local conditions, such as the mountains and canyons, and how they affect winds, which are crucial to how fast and how far a fire burns. During some large fires in the western U.S., a fire weather specialist is sometimes dispatched to the scene to keep firefighters informed of changing conditions.

Keywords to the Wise

You hear these words all the time on your local forecast: highs and lows and temperatures and pressures and wind chills and humidities and chances of showers and wind speeds and directions. But just how slight or how likely is a chance of showers? What exactly are you being told?

Precipitation

When a forecaster sees rain or snow on the way, she has some decisions to make about how to describe the precipitation she expects. She needs to think about how likely it is, how long it will last, and how intense it will be. What is the *probability,* or chance, that it will rain? Will it rain "'til the cows come home" or only briefly? When it rains, will it "come down like cats and dogs" or merely sprinkle?

Sometimes when people hear the word *rain* in a forecast, that's the only word that catches their attention. They make the mistake of assuming that it is absolutely certain that rain will fall on their head sometime during the day. And if it doesn't, well, "Wrong again, she said it would rain." More likely, the forecaster used a few key words that didn't make the same impression as "*rain* in the forecast."

The National Weather Service has issued guidelines for expressing the likelihood of precipitation in hopes that certain word usages would become standard among forecasters. It hasn't worked out very well.

For one thing, the terms the agency uses are easy to misunderstand. It leans heavily on using something called probability of precipitation, which sounds fairly simple when you hear it. But it can mean one thing for one person and something else for another and might even be used to express different things under different weather circumstances.

Technically, *probability of precipitation* is the chance, expressed as a percentage, that a measurable amount of rain — .01 of an inch or more — will fall someplace in the forecast area in the next 12 hours. It is a combination of two figures — the likelihood of precipitation in the area and the percentage of the area that is likely to get it. This doesn't tell you as much as you might think. If there is a 50 percent chance that rain will develop anywhere in the forecast area, and 40 percent of the area is expected to get rain, this translates into a 20 percent chance of rain.

A 70 percent chance of rain doesn't tell you that there is a 70 percent chance of rain everywhere in the forecast area, for example, and it doesn't tell you that rain will fall on 70 percent of the area. It's easy to think that a 70 percent chance of rain means a pretty hefty rainfall is likely, and a 20 percent chance means that only a little rainfall is on the way. But that would be wrong too. Probability of precipitation doesn't tell you anything about how long it is going to rain or how much it is likely to rain.

It's easy to see why many television meteorologists put a different spin on the forecast, finding other ways to tell viewers what to expect. Other meteorologists at private forecasting firms such as AccuWeather Inc. employ a more user-friendly "public communications meteorology" that is intended to better describe how the day's weather is likely to go than the National Weather Service's way of saying things.

The National Weather Service forecast might say something like this: "Partly sunny today with showers and thunderstorms likely in the afternoon. Probability of precipitation is 70 percent."

AccuWeather's version of such a forecast might go something like this: "Sunshine this morning, clouds will billow up during the afternoon, bringing much of the area a brief downpour between 5 and 8 p.m."

Often, you will hear forecasters use a variety of key words to give more meaning to the probability statistics that the computer models put out. These words can describe the likelihood of rain in other ways. They might tell you how certain the chances are of rain, for example, or they could tell you how much area the rain is likely to cover.

A slight chance of rain, for example, generally means that the probability of rain is 20 percent or less. Expressed another way, a slight chance of rain can be described as scattered showers. At least, that's what the National Weather Service has in mind.

When a forecaster says there is a 30 percent chance of showers, it is worth keeping in mind the other side of the coin: There is a 70 percent chance that it won't rain! When a forecaster predicts scattered showers, it doesn't mean it is going to scatter showers on you. (So count yourself lucky and give your forecaster a break!)

Another set of four key words often is used to describe the duration of the rain that forecasters expect: *brief, occasional, intermittent, frequent*.

A weather forecaster also can use key words to describe the intensity of the rain to expect. Nobody does it exactly the same way, but here are how some of the key words are commonly used:

- *Very light*, when less than .01 of an inch is expected
- *Light*, from .01 to .10 inch per hour
- *Moderate*, from .10 to .30 inch per hour
- *Heavy*, for anything above .30 inch per hour.

Temperature

Daytime maximum temperatures and nighttime minimums that a forecaster expects are often expressed as ranges. If the forecast covers a wide area, the highs and lows are not going to be the same in one place as in another. While few forecasters would pretend that they can predict the exact degree of high and low temperatures, still single numbers might be favored. Studies have shown that single numbers are easier to remember than ranges of numbers or words that describe ranges.

Temperatures of the air are very sensitive to local conditions, such as the presence of a tree or a parking lot. The tree will keep you cooler during the day, of course, and the parking lot is hot. But these features of the landscape may have the opposite effect after dark. You may notice that the tree may actually have a warming influence on the temperature overnight, while the exposed parking lot can be especially cold.

Temperature ranges are generally forecast in units of five degrees, and you commonly hear these key words used to describe them:

- *Near,* as in near 45
- *Around,* as in around 20 degrees
- *About,* as in about 75
- *Lower,* as in lower 60s (60–64)
- *Mid,* as in mid 60s (63–67)
- *Upper,* as in upper 60s (66–69)

Forecasts of conditions beyond a couple of days are likely to use wider ranges of temperatures, calling for temperatures, for example, simply "in the 60s."

Short-term forecasts for a specific area often are expressed in specific numbers, such as 70/52/74 to indicate a high of 70 degrees, an overnight low of 52, and a following daytime high of 74.

To avoid confusion, when temperatures get below 10 degrees or above 100, specific numbers are often used, even to express ranges. The forecast might be for a high of *near* 106, for example, or 102 to 108.

Temperature's relative humidity

What a funny country. Everybody east of the Rocky Mountains is very familiar with the idea of *humidity,* the amount of water vapor in the air, and a lot of people west of the Rockies don't know what all the fuss is about. (Chapter 6 explains these different climates.) Out West, when the temperatures get uncomfortably high, people console themselves with the idea that "It's a dry heat."

When it comes to the human body's comfort zone, the temperature of the air around it is only part of the story. A summer day when the maximum temperature hits 95 degrees in Sacramento, California, for example, is a very different experience than a summer day when it reaches 95 degrees in Chicago or Pittsburgh. In Sacramento, you might hear people remarking about the "nice weather." In Chicago or Pittsburgh, people are starting to get the look of disaster on their faces.

The moisture of the air — the amount of water vapor it contains — is important to how you and I feel about the temperature of the air and important to how the air behaves when it comes to making weather. Moisture determines how much more water vapor it is likely to absorb through evaporation, on one hand, and how likely it is to give up water vapor through condensation, on the other. These tendencies are often measured by a property called *relative humidity*.

Relative humidity is a little tricky. It is expressed as a percentage, which is easy to get, well, not quite right. When a weather forecaster says the relative humidity is 40 percent, for example, he is not saying that the air contains 40 percent water, or even 40 percent water vapor. *Relative humidity* describes the percentage of water vapor in the air in relation to the total amount of water vapor the air can contain at that temperature. The air is 40 percent along the scale between holding absolutely no water vapor (which never happens, by the way) and holding all that it can, its point of *saturation,* which happens a lot.

Water vapor is an invisible gas, but you see and feel the effects of the air's relative humidity all of the time. When you take a shower, for example, you're adding so much water vapor to the air that it is quickly saturated. Condensation is producing tiny water droplets and you're standing in a little cloud. Step out of the shower, and you might begin to feel a little chill. You are stepping into air that is less humid, and heat is leaving your wet skin through evaporation. The heat is used to convert the liquid into the water vapor that is being absorbed into the air. The bathroom mirror is fogged up because the air up against its cooler surface has fallen to its saturation point and some of its water vapor has condensed into dew.

Clouds form when air rises and cools to its point of saturation.

Relative humidity on a warm day can make all the difference in how you feel. When the relative humidity of the 95-degree air is low, it is relatively easy for the body to cool itself. Perspiration seeps through the pores in the skin, and the moisture easily evaporates into the air. Heat energy burned up in changing the water from liquid to vapor leaves the body's temperature cooler. When the relative humidity of the 95-degree air is high, however, it really doesn't want to take up much more water vapor, thank you, so it is much harder for the body to get rid of its heat and its perspiration. So the heat builds up, the perspiration accumulates, and the poor body, well, it just kind of stews.

Fahrenheit and Celsius

The world of weather has two main ways of measuring the temperature of the air (and the temperature of water, for that matter). The Fahrenheit scale is used in the United States; the Celsius scale is used in much of the rest of the world, including Europe and Canada.

In 1714, German physicist Gabriel Daniel Fahrenheit developed a scale where water boils at 212 degrees (F) and freezes at 32 degrees (F).

In 1742, Swedish astronomer Anders Celsius devised a scale in which the melting point of ice is designated zero degrees (C), and the boiling point of water is 100 degrees (C).

This figure is a handy comparison of the two scales.

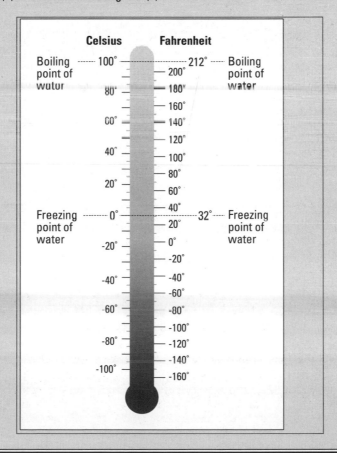

Heat index

The National Weather Service's Heat Index, shown in Figure 1-1, shows *apparent temperature* values — the temperature that the body actually *feels* — which is more important than simply the heat or the humidity of the air. It is not just the heat or just the humidity, but both of these properties that this index combines to more accurately reflect what your body is going to have to cope with. The numbers in the index are based on shady, light wind conditions. Full sunshine exposure can raise these numbers by as much as 15 degrees.

Figure 1-1:
This Heat Index was devised by the National Weather Service to show the temperature the body feels when the heat and humidity are combined.

Apparent Temperature Index

Air temperature (°F)	Relative humidity (%)																				
	0	5	10	15	20	25	30	35	40	45	50	55	60	65	70	75	80	85	90	95	100
120	107	111	116	123	130	139	148														
115	103	107	111	115	120	127	135	143	151												
110	99	102	105	108	112	117	123	130	137	143	150										
105	95	97	100	102	105	109	113	118	123	129	135	142	149								
100	91	93	95	97	99	101	104	107	110	115	120	126	132	138	144						
95	87	88	90	91	93	94	96	98	101	104	107	110	114	119	124	130	136				
90	83	84	85	86	87	88	90	91	93	95	96	98	100	102	106	109	113	117	122		
85	76	79	80	81	82	83	84	85	86	87	88	89	90	91	93	95	97	99	102	105	108
80	73	74	75	76	77	77	78	79	79	80	81	81	82	83	85	86	86	87	88	89	91
75	69	69	70	71	72	72	73	73	74	74	75	75	76	76	77	77	78	78	79	79	80
70	64	64	65	65	66	66	67	67	68	68	69	69	70	70	70	70	71	71	71	71	72

Here are some valuable health guidelines from the National Weather Service:

- ✔ Between 80 degrees and 90 degrees on the Heat Index, be on the lookout for signs of fatigue with prolonged exposure or physical activity.

- ✔ Between 90 degrees and 105 degrees, with prolonged exposure or physical activity, you *may* experience sunstroke, heat cramps, or heat exhaustion.

- ✔ Between 105 degrees and 130 degrees, you *are likely to* experience sunstroke, heat cramps, or heat exhaustion.

- ✔ At 130 degrees or above, dangerous conditions of heat stroke or sunstroke are *highly likely* with continued exposure.

Wind

Winds have a lot to do with how storms come and go (Chapter 4 goes into detail), but the weather forecast concerns itself mainly with the wind in your face. A forecaster describes what is expected of the wind's direction and its speed.

Wind direction describes where the wind is blowing *from*. And so a north wind is coming out of the north and blowing toward the south. You get the idea. Weather forecasts commonly describe the direction of winds on an eight-point compass: north, northeast, east, southeast, and so on.

Winds that blow only up to 5 miles per hour are generally described as *light*, or *light and variable*, to indicate that they are kind of wafting around in different directions. Winds 15 to 25 miles an hour sometimes are described as *breezy* when its mild weather or *brisk* when it's cold. The word for 20 to 30 miles per hour is usually plain *windy*, at 30 to 40 miles per hour they are *very windy*, and winds blowing 40 miles per hour or greater can be described as *strong, damaging, dangerous,* or *high*. Winds become "hurricane force" at 74 miles per hour, but they are dangerous well before then. (Stay out of them!)

Wind chill

Wind can make hot temperatures feel cooler and cool temperatures feel colder.

This is the *wind chill* factor — how a wind makes the body feel that the air around it is colder. Wind chill can be an important indicator of the danger of severe cold winter conditions.

The warmth of the body actually creates a thin envelope of warm air around it, a little insulating comfort zone. Along comes the wind that whisks that envelope away, exposing the skin to the raw cold and accelerating the heat lost by the body. The stiffer the wind, the greater the heat loss, the colder the feeling of air.

Figure 1-2 is a handy index from the National Weather Service showing the effect of wind on the body's sense of temperature.

Figure 1-2:
This official Wind Chill Index shows the effect of wind speed, the left column, in miles per hour, on the body's sense of temperatures.

Wind-chill equivalent temperature.

Wind speed (mi/hr)	Air temperature (°F)																
	35	30	25	20	15	10	5	0	-5	-10	-15	-20	-25	-30	-35	-40	-45
5	32	27	22	16	11	6	0	-5	-10	-15	-21	-31	-31	-36	-42	-47	-52
10	22	16	10	3	-3	-9	-15	-22	-27	-34	-40	-52	-52	-58	-64	-71	-77
15	16	9	2	-5	-11	-18	-25	-31	-38	-45	-51	-65	-65	-72	-78	-85	-92
20	12	4	-3	-10	-17	-24	-31	-39	-46	-53	-60	-74	-74	-81	-88	-95	-103
25	8	1	-7	-15	-22	-29	-36	-44	-51	-59	-66	-81	-81	-88	-96	-103	-110
30	6	-2	-10	-18	-25	-33	-41	-49	-56	-64	-71	-86	-86	-93	-101	-109	-116
35	4	-4	-12	-20	-27	-35	-43	-52	-58	-67	-74	-89	-89	-97	-105	-113	-120
40	3	-5	-13	-21	-29	-37	-45	-53	-60	-69	-76	-92	-92	-100	-107	-115	-123
45	2	-6	-14	-22	-30	-38	-46	-54	-62	-70	-78	-93	-93	-102	-109	-117	-125

Sky cover

TIP

Clouds are a big part of weather, of course, and Chapter 5 is all about clouds. The extent of cloudiness has a lot to do with how the day feels, and the forecaster can use a variety of words to describe this big feature of weather. While different forecasters have their own ways of describing what to expect up there, the National Weather Service uses these terms to describe the extent of cloud coverage of the sky.

- ✔ When the sky will be free of all cloudiness or less than one-tenth cloudy, forecasters describe it as *clear* or *sunny*.

- ✔ At three-tenths to six-tenths, the sky is *partly cloudy* or *partly sunny* or *scattered* with clouds.

- ✔ At seven-tenths to eight-tenths, the sky is *mostly cloudy* or cloudiness is *broken*.

- ✔ At nine-tenths to the whole shebang, it's just plain *cloudy* or *overcast*.

Tools of the Trade

BIG IDEA

For all the difficulties in forecasting its behavior, just a few key properties of the atmosphere are pretty much responsible for the weather it produces and relate to the amount of energy up there. One is the temperature of the air, which gives a good idea of the amount of energy in it. Another is its moisture content, or *humidity,* which gives a good idea of how much rain or snow it could produce. And the third most important property is the air pressure it exerts. This controls its horizontal motion — what you call wind — and influences its vertical movement up and down through the atmosphere. Another key factor is winds. (Chapter 4 discusses the details of air pressure and winds.)

Among the important tools of the trade for modern weather forecasting are the instruments that measure these key properties of temperature, moisture, and motion. Another set of tools you might think of are instrument carriers that transport the measuring devices into various realms of the atmosphere. As tools of modern weather forecasting, computers are in a category by themselves.

The instruments

Thermometers measure air temperature. They do this by registering the air's heat content. In traditional thermometers, the heat causes their liquids to expand and contract, traveling up and down their tubes as temperatures rise and fall. These days, electronic thermometers and other instruments are widely used. Official thermometers measure air four to six feet from the ground in a shelter that is built and maintained to rigid specifications.

Hygrometers or *hydrometers* are instruments that measure humidity, the amount of water vapor in the air. The moisture content of the air not only affects the comfort — or discomfort — you feel at different temperatures, it is an indicator of the chances of clouds, fog, or rain as well.

Barometers measure air pressure by weighing the amount of air over a particular spot. The importance of the barometer is not that its air pressure measurements are particularly noticeable like temperature or humidity. A barometer is an important weather predictor. Falling pressure often means that storminess is coming, and rising pressure usually points to improving weather.

Wind direction is indicated by a *weathervane*, one of the oldest weather instruments. An arrow or some other figure swivels on rod in line with the wind. Wind speed is measured by an *anemometer*, a spinning wheel of cup-shaped blades, which capture the strength of the air's motion. Standard wind measurements are made at a height of 33 feet. At small airports, *windsocks* are tube-shaped flags that give pilots quick visual evidence of the direction and strength of wind.

Rainfall and snowfall are tricky to measure, because winds can cause precipitation to be deposited unevenly, pooling or stacking up a little more here, a little less there. *Rain gauges* are straight-sided tubes or buckets that weather observers place in relatively open spaces near the ground. A *snow gauge* is a similar device that has a flat surface that is periodically cleared off during a storm to prevent readings of snow amounts from being distorted by melting or the snow's compaction from its own weight. A large national network of cooperative observers periodically measures snow depths that accumulate through the winter.

The instrument carriers

Accurate weather predictions depend on the numerical models being able to know what the atmosphere is doing in places that are hard to get to. They need to know the temperature, humidity, air pressure, and wind speeds and directions not only at the surface, but also high in the sky, in the upper atmosphere. They need to know what is going on not just around the corner or over the hill, but in remote areas and also far out at sea. Thousands of people and thousands of lonely pieces of automated equipment around the world help fill these needs every day. The information constantly streams into national data centers around the world, and from there it goes into the supercomputers running the Numerical Weather Prediction forecast models.

Reading the instruments that measure temperature, humidity, air pressure, and winds is a big item for the 179 member nations of the World Meteorological Organization, which operates under the auspices of the United Nations. This chapter's earlier section, "Take what is happening now . . . ," goes into the details of why these observations are so important to weather forecasting. The WMO collects and distributes data from 10,000 land-based observing stations, from 7,000 stations aboard ships, from automatic weather sensors on buoys in the sea, attached to balloons, and on aircraft and satellites.

Weather balloons

Weather balloons filled with hydrogen gas are released from 1,000 sites, including about 80 in the United States and Canada. They are released every day, twice daily, at the same hours, which correspond to 7 p.m. and 7 a.m. Eastern Standard Time, as shown in Figure 1-3. Readings from instruments carried by these balloons are transmitted to ground recording stations. Balloons are tracked by radar to help measure wind speed and direction in the upper atmosphere.

After about two hours of flight, the average balloon reaches an altitude of about 20 miles. About that time, the balloon bursts, and the instrument package parachutes back to the surface. Many are lost over the ocean or in remote land areas, although often people find them and, following the instructions on the packages, return them to the National Weather Service to be used again.

National Center for Atmospheric Research/University Corporation
for Atmospheric Research/National Science Foundation.

Figure 1-3:
A National
Weather
Service
forecaster
releases a
weather
balloon.

Ocean buoys

Ocean buoys, shown in Figure 1-4, deliver instrument readings of ocean water temperature, as well as standard air temperature, humidity, and pressure measurements, from key locations at sea. Automatic weather sensors are attached to 300 buoys that are permanently moored or drift with ocean currents. Radio signals carry the data from these ocean-based sensors.

A special "Tropical Ocean Array" network of 70 buoys is moored in water up to 15,000 feet deep across the Pacific Ocean at the equator so that weather scientists can keep track of ocean temperature changes that affect weather patterns and seasonal differences in the United States and elsewhere. Across the tropical Pacific is where El Niño and La Niña conditions develop. Chapter 6 describes El Niño and La Niña and goes into the details of how the ocean and the atmosphere work together.

National Oceanic and Atmospheric Administration/Department of Commerce.

Figure 1-4:
Researchers
service the
instruments
on a moored
ocean buoy.

Automated weather stations

Across the United States are about 1,000 stations of the Automated Surface Observing System, each measuring surface weather conditions as often as every minute. Operating automatically, 24 hours a day, these stations acquire and distribute information on cloud cover, visibility, temperature, air pressure, wind speed and direction, and precipitation.

Doppler radar

A powerful remote-sensing tool, Doppler radar is especially good at detecting hazardous thunderstorms that other instruments and forecasting methods find hard to predict or observe. (Did you know that *radar* stands for *ra*dio *d*etection *a*nd *r*anging?)

Conventional radar sends out short, powerful microwave pulses. A receiver picks up the signals that bounce back as the radar waves encounter objects, such as precipitation particles. It gives a rough description of the size of the object and an estimate of its distance.

Doppler radar measures frequency differences between signals bouncing off objects moving away or moving toward its antenna. (Austrian physicist Christian Doppler first described this shifting effect that the direction of travel has on waves of light and sound.) Its ability to detect the Doppler shift in signals allows Doppler radar to paint a detailed picture of the motion of winds inside thunderstorms. (Chapter 9 goes into these storms in detail.)

A network of 146 of these Doppler radars exists around the United States in the network maintained by the National Weather Service, the Federal Aviation Administration, and the Department of Defense. Hundreds more are owned by private companies, including television stations and utilities.

Shaped like a giant ball on top of a scaffold, a Doppler radar emits three billion microwave bursts each second. Their antennae are so sensitive to the microwaves reflected back that they can pick up precipitation, dust, insects, and even differences in air density.

A fleet of satellites

Imagine what it must have been like to try to accurately predict the arrival of a storm from the sea — the Pacific Ocean, for example, or the Gulf of Mexico — before satellites were up there taking great big pictures of the earth. Before 1960 and the launch of Tiros I, about all that a forecaster had to work with were occasional radio reports of a few aircraft or ships at sea. Many severe storms went undetected until they were dangerously close to people. The Galveston Hurricane in 1900, which Chapter 8 describes, is a tragic example.

The pictures sent by weather satellites since those days not only have improved the skill of weather forecasting, they have changed the way a lot of people think about the weather. It was as if the true scope and character of weather events suddenly dawned on everybody at once. Before weather satellites, it wasn't so obvious to everyone that many weather systems are enormous, global-scale features that travel thousands of miles over periods of several days.

These days, of course, a whole fleet of satellites is up there keeping their eyes on things.

Stationary workhorses

Most of those great animated pictures of moving clouds that are the bread and butter of modern weather news on U.S. television come from one of two _geostationary satellites_ parked over the eastern Pacific and western Atlantic oceans. These two workhorses are part of a network of five geostationary satellites positioned around the world at an altitude of 22,238 miles above the equator. They are called _geostationary_ because their west-to-east motion matches the Earth's rate of rotation, which keeps them over the same spot on Earth.

One of the great things about these satellites is that they send back to Earth their pictures in *real time,* as soon as they are taken, so forecasters get up-to-the-minute images of what's going on.

In addition to those pictures of sunlight bouncing off of clouds, these satellites take infrared images that can track storm movements through darkness and help forecasters see that some clouds are higher or thicker than others. Another instrument allows satellites to detect water vapor in the atmosphere, information that is especially valuable to forecasters on the lookout for dangerous storms.

Polar orbiters

At an orbit only about 450 miles up, two satellites are continuously monitoring Earth from the North Pole to the South Pole. These polar orbiters get a much closer view of things.

As they travel around from north to south and back again, the Earth rotates under them, and so their instruments scan a different swath of Earth farther to the west with each orbit. Each satellite covers the entire Earth in 12 hours. Because of their sensitive instruments, these satellites are able to provide forecasters with more detailed pictures of individual storms.

And more . . .

A whole range of exotic detection devices are aboard other satellites designed by researchers who are using their ingenuity to more accurately measure the features of weather and climate.

One joint U.S.-Japan satellite, for example, focuses its imagers and instruments entirely on the Tropics — the belt around the Equator from 35 degrees north and 35 degrees south — where the rain accounts for two-thirds of the rainfall on earth. The storms in this big, warm, and watery midsection pump most of the moisture that travels as water vapor through the atmosphere.

A U.S.-France satellite measures ocean temperatures by detecting slight differences in sea level. This kind of indirect measurement works because a column of warm water is higher than a column of cool water. Data from this satellite, called TOPEX-Poseidon, is especially useful in keeping track of weather-altering climate patterns known as El Niño and La Niña, which Chapter 9 describes in detail.

Another satellite is indirectly measuring the speed and direction of winds on the surface of the seas by detecting minor differences in the reflected microwave "scatter" patterns off of ripples and waves.

A for accuracy, E for effort

The National Weather Service has spent a bundle — $4.5 billion — in recent years to modernize its equipment and improve its forecasting skill, and it seems to be paying dividends.

The National Weather Service says that today's three-to-four-day forecast is as accurate as the two-day forecast was 15 years ago. Predictions of rain three days ahead of time are as accurate today as one-day forecasts were in the mid-1980s. The three-to-five-day long-range forecasts that have been the standard will soon be replaced by seven-day to as much as ten-day predictions.

The accuracy of flash-flood forecasts has improved from 60 percent correct to 86 percent during the past three years. At the same time, potential victims of flash floods get a 53-minute warning now instead of less than eight minutes in 1986.

And the "lead times" of advance warnings of tornadoes — the time that residents have to react to these emergencies — have increased from an average of about five minutes in 1986 to an average of 12 minutes in 1998. Severe local thunderstorms and similar cloudbursts are typically seen 18 minutes beforehand now, up from 12 minutes a decade ago.

Computers

If you haven't thought of a computer as an important weather forecasting tool before, now it's time that you did! Computers are crucial not only in the manipulation of tons of weather information, but also in efficiently communicating this mass of digital data around the world.

The mainstay of modern forecasting, the numerical weather prediction process I describe in this chapter would not be possible without high-powered computing. The National Center for Atmospheric Research in Boulder, Colorado, uses a supercomputer that can perform 2.2 billion calculations per second. Still, researchers say that they could use even more computing power to make their forecast models more accurate and to produce them more quickly.

It's not just the number-crunching power of computers that is important for weather forecasting. Their ability to quickly prepare easy-to-grasp graphical pictures of their output is especially helpful to forecasters. A new advanced system of computers installed in all National Weather Service forecast offices across the country allows forecasters to easily compare these models by running their forecasts simultaneously on the same computer screen.

During dangerous or hazardous storms, when a forecaster's time is critical, computers now take over the job of composing and sending public emergency warnings. Even the NOAA Weather Radio alerts are spoken by computer.

These automated announcements have a certain, well, digital quality to them, and people complain about the use of this automated equipment to issue public warnings in life-threatening circumstances.

How to Read a Weather Map

Any forecaster will happily tell you, computers make drawing a weather map a snap! Reading a weather map has become a lot easier, too.

Weather maps have been around for the better part of two centuries, and for much of that time, they were the only way to visualize a lot of what was going on — or what forecasters thought was going on — in the atmosphere. But even now, when satellite images of actual storms and fronts are available, often a television forecaster and a newspaper weather page will display a simplified weather map to make clear to viewers and readers what's going on.

The typical stripped-down version of a weather map highlights a few features of the weather picture across the nation. As Figure 1-5 illustrates, centers of high pressure and low pressure are designated simply by a large "H" or "L" on the map. Cold fronts lead with arrows ahead of them and warm fronts with semicircles along their leading edge. Areas of precipitation are often designated with hatch-marks or shaded sections. Some temperature readings often are included, and occasional arrows will point out the directions of surface winds. Chapter 2 goes into more detail about these weather features.

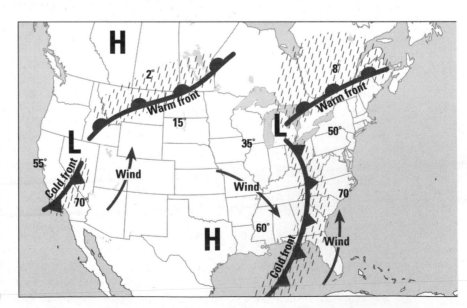

Figure 1-5:
A typical weather map showing features and symbols that are common to most simplified maps used in televised forecasts today.

Chapter 2

Behind the Air Wars

. . . .

In This Chapter

▶ Watching air masses amass and their fronts do battle

▶ Seeing how the Sun's energy powers the weather

▶ Sorting out the effects of orbit and tilt and rotation

▶ Composing an atmosphere

. . . .

If there is a personality that describes the atmosphere, the blanket of air where all of the weather takes place, you might think of it as flighty or fickle. Do you know somebody who changes his mind a lot? Who often seems to be repeating the ideas of the last person he was with? That's the atmosphere all over. (You want to scream sometimes: "Make up your mind!" Does it help?)

At the Go Figure Academy of Sciences (GoFAS), weather experts describe this maddening characteristic in polite, five-dollar terms like *instability* and *turbulence* and *chaos*. Frankly, the word *unbalanced* comes to mind, if you get my drift.

As any weather forecaster can tell you, the atmosphere is unreliable — here today, gone tomorrow, as the saying goes, blowing hot and cold. You think you know it when you go to bed at night, and then, poof! — as soon as the Sun comes up, there's a completely different character. This chapter is all about the things that make the atmosphere and its weather the way it is — so changeable.

I Don't Like Your Latitude!

Don't be too hard on the atmosphere. (After all, it's the air you breathe. Without it, you're sunk.) Imagine trying to live in a house where it's always hot at one end and always cold at the other. That's the situation earth's atmosphere finds itself in. You don't have to be an Eskimo or a Pacific Islander to figure out where the warm spots and the cold spots are, but it helps.

The low latitudes, around the Equator, get a lot of warm sunshine, and the high latitudes, around the poles, get very little. (Figure 2-1 shows the layout of the imaginary lines around the Earth called latitudes.) This temperature difference is no small matter. On the same day, it can reach 120 degrees below zero overnight in Antarctica and 120 degrees above zero in a subtropical desert. The atmosphere has to deal with these huge temperature differences all the time. They have a lot to do with what weather is all about.

You might think things would be pretty comfortable in the middle of the house, the middle latitudes, where most people live in the world. Not too hot, like Goldilocks said, and not too cold. And certainly it's true that, on average, the mid-latitudes are less extreme environments to live in. But there's a catch, of course. In case you haven't noticed, it is in the middle latitudes where you and I live that a lot of the masses of northbound tropical warm air and southbound cold polar air come together. And when they do, it's not a pretty picture. In fact, it can get messy.

Figure 2-1:
Imaginary lines called latitudes divide the world into the Tropics, the polar regions, and between them, the mid-latitudes.

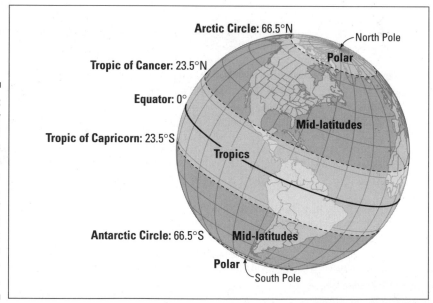

Arctic Circle: 66.5°N
North Pole
Polar
Tropic of Cancer: 23.5°N
Equator: 0°
Mid-latitudes
Tropic of Capricorn: 23.5°S
Tropics
Antarctic Circle: 66.5°S
Mid-latitudes
Polar
South Pole

Those big storms with the long lines that twist together into a whirling low pressure system on the weather map are called *mid-latitude cyclones,* or sometimes, *extratropical cyclones.* You might think of these storms as battles between opposing armies of hot air and cold air. The lines between them are very appropriately referred to as *fronts* — as in battlefronts. Weather scientists used to think of these storms exactly this way, although now the picture in their minds is more complicated. (The section later in this chapter, "News from the fronts," goes into more detail about these clashes.) Nobody who has lived through the devastation of a flood or a blizzard or an especially severe winter needs to be reminded that weather unleashes powerful forces that can do terrible things. You and I are bystanders to all of this, and the only thing to do when a battle is raging is try to stay safe and warm and dry.

Where the Armies Mass

An *air mass* is a widespread body of air that has a uniform look to it. Across hundreds if not thousands of miles, all this air looks and acts pretty much alike. This is because it has been parked over a particular region of the earth long enough to absorb some of its important qualities. It has picked up from the surface a certain temperature and humidity, or moisture content, and like in any good army, these characteristics are fairly evenly distributed. There are no real storms or battles going on or even strong winds — this is all the same army, after all — and pressure, like morale, is fairly high.

Just as different regions of the world have certain characteristics, so do the air masses that form over them. The big differences that make dramatic weather are temperature and humidity — they are warm or cold and moist or dry.

- *Continental* air masses form over land and are dry.

- *Maritime* air masses form over an ocean and are moist.

- *Polar* or *Arctic* air masses are cold.

- *Tropical* air masses are warm.

A weather war zone

Every place has its own dangerous weather at one time of the year or another. But did you know that the continental United States gets more violent weather than anyplace else on earth? This came as a big surprise to early American settlers.

In a typical year, the National Weather Service reports, the continental United States can expect these violent weather events:

- Roughly 10,000 severe thunderstorms

- About 1,000 tornadoes

- Approximately ten severe winter storms

- An estimated 1,000 flash floods

- Plus threats most years from tropical storms and hurricanes in the North Atlantic, the Gulf of Mexico, and the Caribbean Sea

Why all these extreme weather events? It's like they say about the real estate business: The three most important things are location, location, and location.

As Figure 2-2 illustrates later in this chapter, one region or another of the continental United States is in pathway of a variety of moist and dry and warm and cold air masses coming at it from all directions.

Lake-Effect snows

In late fall and early winter, people who live along the southern and eastern shores of the Great Lakes are very familiar with the effects of big expanses of water on the air. The continental polar air mass that brings cold, clear winter days to the Midwest has a very different look to it by the time it crosses the Great Lakes.

Even after — or if — the lakes freeze over, still they affect the amount of snowfall in the region. Four main weather-generating processes are at work when the cold, dry Canadian air flows over the lakes and the regions along the southern and eastern shorelines.

✔ The dry air absorbs moisture from the lake, which fuels the the snowfall and gives the air more latent heat energy that is released as it condenses into a cloud.

✔ Flowing over the warmer lake water, the temperature of the cold air rises, and as it warms, the air lifts higher into the sky and becomes more unstable.

✔ After blowing over the smooth surface of the lake, the air plows into the uneven surface of the land, slows down in the friction, and piles up, rising farther into the sky.

✔ And the cold air gets another lift from the hills and higher terrain it encounters above the far shores.

In the summer, the Great Lakes have a very different effect on the weather of the region. Because the water of the lakes is often cooler than the surrounding air this time of year, they can dampen thunderstorm activity.

The action begins when an air mass moves from its place of origin, usually in response to winds in the upper atmosphere. While weather scientists now think of the winds high in the atmosphere as the driving forces behind the storms, the weather in your face still has the look and feel of a battle of air masses. In the *Northern Hemisphere,* the half of the world north of the Equator, you can be sure that a cold air mass is moving down out of the cooler regions of the north and a warm air mass is moving up from the warmer regions in the south.

Winter air masses

Continental polar air masses that move down out of the snow-covered regions of northern Canada and Alaska are often in the picture when bitterly cold winter weather visits the United States. Winds of this frigid, dry air blow over the northern plains and through the Midwest and the Northeast and occasionally reach as far south as Texas and Florida. Out west, the Rockies, Sierra Nevada, and Cascade Mountains usually protect the Pacific Coast. As these air masses move southward, their extreme cold tends to modify, protecting the southern states from the most extreme cold. The barrier of the Appalachians sometimes protects the cities of the East Coast from the worst of it.

Figure 2-2: The map shows the different air masses that affect weather in the continental United States and helps explain why the nation gets so much dramatic weather.

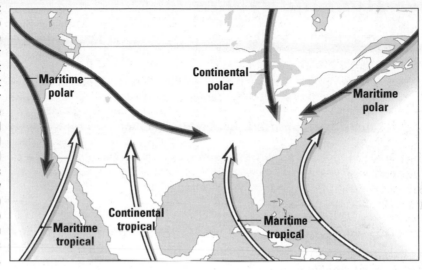

Maritime polar air masses regularly sweep over the West Coast from out of the northern Pacific Ocean, bringing cool, moist air that dumps heavy snow in the western mountains. By the time this air crosses the Rockies, it is relatively dry. It warms as it blows down the eastern slopes of the mountains and sometimes brings fair weather and warming temperatures to the plains. Maritime polar air also originates in the North Atlantic and occasionally sweeps southwestward into New England and the mid-Atlantic states, bringing rain and snow.

Maritime tropical air masses from the Pacific sometimes sweep northeastward over California, and when they do, they can bring heavy rains and flooding. Similarly, *maritime tropical* air from out of the Gulf of Mexico and Caribbean Sea can fuel flooding storms as it moves up the Mississippi and Ohio valleys.

Summer air masses

The *continental polar* air that brought frigid misery during winter often has a welcome cooling effect over much of the United States in summer. Fair weather generally prevails over the Northern Plains and Midwest when continental polar air is around, occasionally reaching even the Gulf of Mexico with pleasant dry air.

Maritime tropical air masses from the Atlantic Ocean, the Caribbean Sea, and the Gulf of Mexico spend a lot of time over the eastern two-thirds of the United States during summer, especially the Southeast, bringing warm temperatures and high humidity.

Continental tropical air forms over the big desert regions of northern Mexico and the U.S. Southwest in summer, keeping things hot and dry.

News from the Fronts

During the most powerful storms of winter and summer you will probably find two air masses of very different qualities — different temperatures and humidities — crashing into one another. When upper winds and other conditions are right, this is where you will see a powerful mid-latitude cyclone with well-defined fronts that bring rain or snow over big sections of the country. In spring and summer, along these fronts can be long lines of severe thunderstorms that can form tornadoes when conditions are right — or wrong!

It might look like a rainy or snowy mess from down below, but a *front,* the transition zone or line between two distinct air masses, has a certain shape or structure to it, depending on what's going on.

Cold fronts

When a *cold front* moves into space occupied by warmer air, it forces the warmer air to rise up and over the leading cold wedge. This is because cold air is denser, or heavier, than warm air and sort of bullies it out of the way. As the warm air rises, it mixes with the cold air and condenses into big clouds. Figure 2-3 shows what happens with the passage of a typical cold front.

Often, these cold fronts are carried along by westerly winds, which tear away the tops of these clouds and carry them out far in advance of the approaching front. The first sign of an advancing cold front can be the appearance of these high clouds, followed quickly by a thickening sky. And then a line of heavy thundershowers develops as the front passes, followed by steadier and gradually lighter rain. A typical cold front moves about 30 miles an hour, often traveling to the south and east.

Warm fronts

When a *warm front* moves into the space occupied by a retreating cold air mass, it develops a completely different precipitation pattern than an advancing cold front would. While a cold front digs in underneath a warm air mass, because its air is denser, the lighter air of a warm front rides up above the cold air it is replacing. In this rising warm air, clouds form far out ahead of the advancing warm front's boundary. Hundreds of miles ahead of the surface edge of a warm front, the cloud cover slowly thickens, finally bringing rain or snow. All this happens long before the surface edge of warm front finally passes. When that happens, the sky is usually clear. A typical warm front can travel about 10 to 20 miles per hour, most often moving north and east.

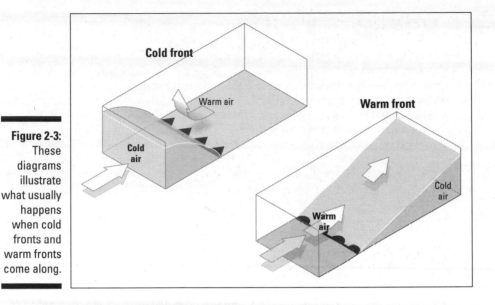

Figure 2-3:
These diagrams illustrate what usually happens when cold fronts and warm fronts come along.

Stationary fronts

A *stationary front,* as you might expect, isn't going anywhere, at least for the time being. But don't let its lack of motion fool you. A stationary or slow-moving front can mean real trouble. A lot of weather can be happening around a front that isn't moving. Sometimes, when warm humid air travels over the frontal boundary, rainfall can occur over large areas, and because the front isn't moving, the rain may stay long enough to cause flooding.

Occluded fronts

As a winter storm progresses, the large wedge of warmer air between the warm front and the cold front gets narrower. The cold front travels faster than the warm front, and beginning near the low pressure center of a mature storm, it eventually overtakes it. This new boundary is called an *occluded front.* On one side is cold air from behind the cold front. On the other is cool air that the warm front was flowing over. Above the front is warm air that has been cut off from the ground.

There goes the moon

Sailors and poets and ancient thinkers all favored the idea that the moon has a lot to do with how things go here on Earth. A "ring around the moon" was seen as a sign of impending storminess, for example, but none of it is the moon's doing.

That ring or halo is moonlight — sunshine reflecting off of the moon's surface — scattering as it encounters thin clouds in the atmosphere. They might be high in the atmosphere, those clouds, but they are nowhere near the moon.

But the moon has this important indirect impact on weather: The tug of its gravity makes the tides of oceans and very large lakes rise and fall, and during powerful coastal storms, high tides can make the difference between inconvenience and natural disaster.

Here Comes the Sun

Weather is the Sun's doing. It might be hard to believe when you are being hammered by a cold winter storm, but the Sun, the star of the solar system, is the driving force behind all this weather, the commander in chief of all these air wars. The heat energy — the solar energy — radiating from this star is the fuel that drives it all, everything I describe in *Weather For Dummies*. If it were not for the Sun's warming of Earth and its atmosphere, the planet would be frozen solid, a ball of ice. (And nobody would be around to read — or write — about the weather.) The Sun's energy is the fuel, and weather is the result of the huge temperature differences between the Equator and the poles. These temperature differences are partly Earth's doing, of course. But if you want something to blame it on, blame the weather on what happens deep within the interior of a star 93 million miles away!

Moving Sun's energy

The atmosphere is always in the process of converting the Sun's energy from one form to another and moving it from place to place. Behind all of the motion, and commotion, is a complicated exchange of energy between the atmosphere and the Earth. Things are out of balance, and the system is trying to even them out. The atmosphere loses more of the Sun's radiation than it absorbs, and Earth's surface has a solar energy surplus. This exchange process is what keeps the atmosphere from becoming unbearably cold and the ground you and I stand on from getting unbearably hot.

Through heat-transfer processes that weather scientists call *conduction* and *convection,* the radiation surplus is continually moving from the surface up into the atmosphere. The heat is conducted directly from the surface to the thin layer of atmosphere above it. You might say that the convection process takes it from there, mixing it through winds and other weather processes among the higher layers of atmosphere.

While weather watchers like you and I focus on the sky, on the rays of the Sun beating down on us and on the rain or snow falling from the clouds, a weather scientist sees things as part of an energy transfer that is moving from the ground up. A winter storm at the seam between two air masses, which this chapter's previous section, "News from the Fronts," is about and Chapter 8 describes more fully, is transferring energy from the surface to the atmosphere. So is a summer thunderstorm, which Chapter 10 describes. And so is an autumn hurricane, which Chapter 7 details. Through conduction and convection, they are all moving one form of heat or another from a warmer region to a cooler one.

Later in this chapter, the section "The Big Picture" explains why the atmosphere never succeeds at this process of balancing the Sun's energy around the world. Figure 2-4 shows what happens to the radiation from the Sun striking Earth.

Reflecting on albedo

Some sunshine is missing. Nobody leave the room.

About half of the solar energy that reaches Earth's atmosphere ends up getting absorbed at the surface. There it gets converted into invisible long-wave radiation. When it re-enters the atmosphere sooner or later, it helps make weather. Another 20 percent gets absorbed by the atmosphere and clouds on the way down.

So what happens to the other 30 percent of the sunshine? People who have been put on the case say it gets lost to *scattering,* when sunlight rays collide with air molecules or tiny dust particles are reflected back off of bright surfaces.

The brighter the surface, the more light it reflects (and the less it absorbs). This is why a white shirt is cooler than a dark one on a summer day. The percentage of light that a surface reflects back, rather than absorbs, is a property that scientists call *albedo.*

Albedo is a big deal. For example, 20 percent of the incoming sunshine bounces right off the bright white cloud tops and heads back toward space. About 4 percent is reflected back from Earth, but there are big differences in the albedo of different surfaces. It ranges anywhere from 95 percent for fresh snow to 2 percent for calm water.

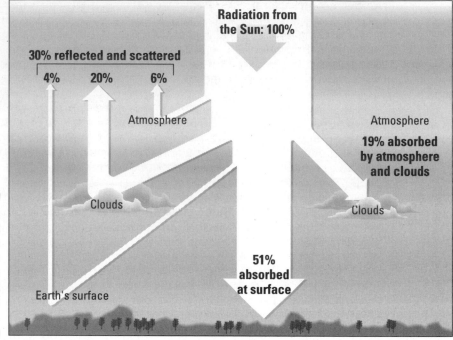

Radiation from the Sun: 100%

30% reflected and scattered

4% **20%** **6%**

Atmosphere

Clouds

Atmosphere

19% absorbed by atmosphere and clouds

Clouds

51% absorbed at surface

Earth's surface

Figure 2-4: Here's what happens to the radiation from the Sun once it reaches the atmosphere.

Looking absolutely radiant!

Don't look now, but waves of energy are radiating all around you. The fact is, everything that has a temperature above absolute zero (-459.67 degrees Fahrenheit) is giving off at least some waves of radiation. Your body, for example. Even your *Weather For Dummies* book. This radiation is an important part of the process of turning the Sun's energy into weather.

Bring in the clouds

Clouds love this long-wave infrared radiation. The tiny water particles in them soak it right up.

Have you ever noticed that a cloudy night often is warmer than a clear one? Those clouds are absorbing that radiation coming off the earth and radiating a lot of it back toward the ground, acting like a big infrared blanket over you.

A cloudy day can be a different story, depending on the season. In spring and summer, the same clouds that kept you warm at night now are preventing the sunlight from reaching you. More often than not, a calm cloudy summer day is cooler than a calm sunny day. In winter, clouds during the day often make temperatures near the ground warmer than on a clear day.

How to cause a storm

Did you know there are two forms of heat?

One is the kind of heat that you feel on your arm, say, when you exercise or hang out in the sunshine. This they call *sensible heat,* because you can sense it. (This sounds pretty sensible to me, although I wouldn't spend too much time out in that sunshine.)

The other kind is *latent heat.* This is heat that is released or absorbed when things like water change *phase,* or form, between vapor and liquid and ice. They call it *latent* because it is stored away, or hidden. (They could have called it *insensible*, you know, but nobody asked me.)

This idea sounds a little tricky at first, but really it's no sweat. Look at it this way:

When you perspire, your body is working on getting rid of excess sensible heat. The sweat on

your arm *evaporates,* converts from liquid water to gaseous vapor. This process of conversion from liquid to gas state absorbs heat, and the coolness you feel is the sensible heat being converted to latent heat. The heat that left your body is stored away in the molecules of that little bubble of air that just lifted off from your arm.

Now follow that water vapor off of your arm as it rises up higher and higher into the sky and forms a cloud. When it does this, it converts itself back into liquid, tiny water droplets, and gives the heat it took off your body back to the atmosphere. Oops, things look a little unstable up there. A storm is brewing! Now see what you've done!

When it comes down from the Sun, however, most of the energy arrives as powerful short-wave radiation, including the spectrum of light that you can see, and this passes right through the atmosphere and strikes the surface of the Earth. Depending on the kind of surface it hits, it bounces back or is absorbed. It all depends on color and surface texture and other properties of the surface. Notice how hot a black asphalt parking lot gets on a summer afternoon, absorbing the heat, and yet, how quickly it cools, or radiates it away, overnight.

People who plan cities and buildings are studying these different heat-absorbing and radiating properties of materials to make downtowns and neighborhoods more energy-efficient and comfortable places to be.

In case you haven't noticed, the energy that radiates back from the surface travels as long infrared waves, which you and I can't see. Invisible it might be, but this form of radiation is more important than you might think. The atmosphere, which lets most of the short-wave sunlight pass right through without absorbing it, catches a lot of the rebounding long-wave heat energy and keeps it around. This produces the so-called greenhouse effect that is discussed in detail in Chapter 12.

This invisible long-wave radiation that is given off by the Sun-warmed Earth is more important to weather than direct sunlight. Somebody who has spent the day in the sunshine may find it hard to believe, but a weather scientist will tell you: The energy radiating back up into the atmosphere from the surface of the Earth as long-wave radiation has more direct effect on weather processes than the short-wave energy that comes directly through the atmosphere as sunshine.

A contagious convection

Heat that is moving from the surface of water or land warms the air just above it in a process of direct transfer known as conduction. This is the way the icy cold of a glass or the boiling heat from a cup travels up the handle of a metal spoon, for example. And just around the metal spoon handle, a thin layer of air is absorbing some of the heat.

From this thin layer of air at the surface, the heat energy finds its way into higher levels of the atmosphere through a process known as *convection,* the vertical mixing of liquid or gas of different temperatures. Convection is what happens when a pot of water boils.

Some of this air mixing happens through the mechanical forcing of wind. This is referred to as *forced convection*. Blowing near the surface, swirling eddies in the flowing air carry the heat up into the sky. Two general rules apply: the faster the wind, the greater this kind of convection. Also, the more uneven the surface — the bigger and more numerous the eddies — the greater this kind of mixing.

Another kind of vertical mixing known as free *convection* depends on buoyancy — the ability of warmer air to rise in cooler air. In the atmosphere, a kind of bubble of warm air is formed near the surface and floats up to higher altitude, above the cooler, denser air around it, much like a hot-air balloon would do. As it rises higher and higher, the bubble of air expands, and as it expands, it cools. This kind of rising and falling of air of different temperatures and densities is going on all of the time.

The process of free convection can be especially noticeable on a warm summer afternoon. The Sun is heating the ground and the heat from the ground is quickly warming the air just above it. Before long, a rising column of warm expanding air is formed. These are the thermal updrafts that soaring birds ride on a warm day.

If conditions are right, if the air bubble contains enough moisture and the cooler surrounding air is unstable, a cloud can eventually form when the rising air gets cold enough for its water vapor to condense into tiny liquid droplets or even ice crystals. (For more about cloud formation, see Chapter 5.) This condensation process gives off still more heat, called *latent heat*. This latent heat plays a major role in the in the formation of clouds and storms. (See the sidebar "How to cause a storm.")

The Big Picture

You want to know what's really behind all that turbulent mess you think of as weather? (No, my people at the Go Figure Academy of Sciences have looked into it, and they tell me it's *not* the government.) Do you want the Big Picture? Well, now it can be told. Believe it or not: *It all has to do with the way the solar system is put together.* (Hey, you wanted the Big Picture!)

When you really get down to it, three large facts of life in this particular reach of the solar system are responsible for the behavior of Earth's atmosphere. You can see them all in Figure 2-5. It has to do with the orbit of the Earth around the Sun. It has to do with the way the planet rotates on its own axis, the way it spins like a top. And it has to do with the fact that it is not spinning straight up and down, but at a tilt. Also, notice how the Earth is closest to the Sun on January 3, in the middle of the Northern Hemisphere's winter. Go figure!

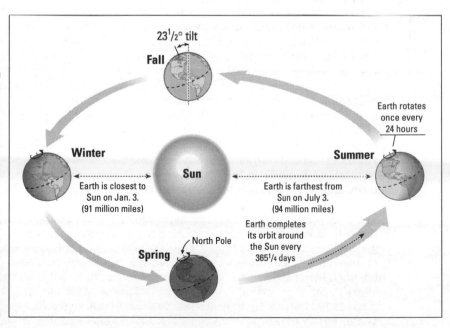

Figure 2-5: The Big Three behind the weather on Earth: its year-long orbit around the Sun, its tilt that gives the year its seasons, and its daily rotation.

Long live the revolution!

If you told me that it takes a year for the Earth to travel completely around the Sun, and that a year is 365 days, you would be accurate enough for most purposes. But I might not want to set my clock by yours. Did you remember Leap Year — the fact that you add a 29th day to February every four years? This makes up for the fact that the complete revolution of Earth's orbit around the Sun actually takes 365¼ days.

There's something else about Earth's orbit of the Sun that is a little, well, irregular. If you look at it closely, you will see that it is not really a circle — that is, the Sun is not in the center of Earth's orbital path. Instead, it is off to one side. The shape of the orbit is elliptical, which means that at some times during the year the Earth is actually *closer* to the Sun than at other times.

This state of affairs might lead you to think — as some people do — that Earth's elliptical orbit is responsible for the fact that some times of year are warmer than others — that summer might be caused by the fact that the Earth and Sun are closest together at that time of year. This is a completely mistaken idea, and you should wash it out of your mind immediately. In fact, I'm sorry I brought it up!

If you have any doubts, consider this fact: The Earth is closest to the Sun every year on January 3. I don't know about you, but I live in California, and while we have had some pretty nice January days, January 3 has never felt much of anything like summer. If I lived in the Southern Hemisphere, south of the Equator, where January 3 is in the summertime, I might think differently about this, of course. But still I would be wrong. Earth being closest to the Sun at that time of year is not responsible for the fact that it is summertime there either.

Here's what it means: On or about January 3, the Earth and the Sun are a mere 91 million miles apart. Six months later, on July 3, at the opposite side of the elliptical orbit, when they are farthest apart, the distance has stretched to 94 million miles. This is about a 3 percent difference in the Earth-Sun distance from one time of year to another.

It has an impact on the intensity of sunshine reaching Earth, no doubt about it. Scientists have figured out that Earth gets 7 percent more heat energy from the Sun on January 3 than it does on July 3. This is because on July 3, even though it is mid-summer in the Northern Hemisphere, the sun's rays are traveling a little farther and so are slightly more spread out than they are on January 3. But this small difference does not account for the seasons. The angle that they strike a particular place on Earth makes a lot more difference to the intensity of the Sun's rays. As Figure 2-6 illustrates, the angle is what the seasons are all about.

Figure 2-6:
In summer, the Sun's rays are more intense as they strike the atmosphere more directly overhead, while in winter they strike at a greater angle and travel through more atmosphere.

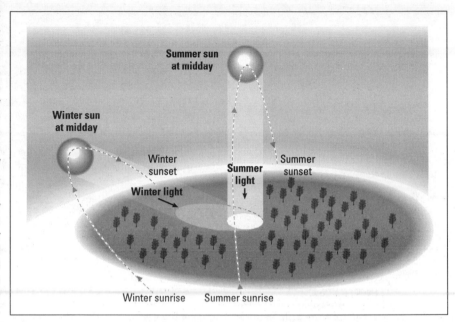

Spreading the beam

The closer together the rays of sunshine, the more intense the energy. This idea helps explain why the energy from the Sun is weaker when it shines on the polar regions of Earth than at the Equator. These regions remain cold even though the Arctic and Antarctica, at the North and South Poles, get many hours of daily sunshine during the summers in the Northern and Southern Hemispheres. It has to do with the angle at which the sunlight strikes.

Try this at home. Notice how much brighter a flashlight's beam is when it is shining directly at a surface and how quickly it fades when you spread the beam out at a greater angle and over a bigger area.

Everybody notices this effect of the sunshine between the different seasons, of course. Unless they live in the Tropics, the region of the world along the Equator, where the Sun is more or less directly overhead all year long. For most of the world, the winter Sun that comes glancing in at a low angle is a pretty weak sister to the summer Sun that spends a lot of time directly overhead. Figure 2-7 illustrates the point. The next section of this chapter explains the cause of this seasonal angle.

The sunlight's angle affects its intensity in another way. The more directly the Sun is over your head, the less of Earth's atmosphere it has to penetrate.

A matter of some gravity

Hmmm . . . let me see now. . . .

The time from the Summer solstice to the Winter solstice would be — yup, that checks out, 182 and 183 days between them, close enough.

And from the spring equinox to the autumnal equinox — oops, what's going on here? Between March 20 and September 22 are 186 days, and between September 22 and March 20 are 179 days.

Sure enough, it has to do with the elliptical shape of Earth's orbit around the Sun (refer to Figure 2-6). This is gravity at work — the pull of the mass of the Sun on the Earth. When the Earth is closer to the Sun, the pull of gravity is stronger. Because it is farther from the Sun from March 20 to September 22, Earth travels more slowly during that loop of its orbit.

This means that summers are seven days longer in the Northern Hemisphere than in the Southern Hemisphere. Do they know this in Australia? Is this legal?

Various things in the atmosphere filter out or scatter some of the incoming rays, so the more atmosphere it has to travel through, the more filtering and scattering takes place. (Chapter 13 says a lot about these optical effects.)

Tilting at the seasons

Earth is out of kilter. You might expect a well-behaved planet to stand up straight, and after 5 billion years or so, to act its age. But nope, not Earth. What can you do? Always it's got this slant to it, like a slouchy teenager, as if it's leaning against something. The angle of this tilt — the difference between where its poles are and where they would be if it were upright in relation to the Sun — is 23.5 degrees. When you come to think of it, this is quite a slant.

This 23½-degree angle is why you have seasons. This is the whole reason why there is winter and spring and summer and fall. This tilt is why there is a time of year when plants are growing vigorously and another time when they are dormant. This slant of the Earth is the reason why January 3 and July 3 have a completely different feel. And this is why some times of year the Sun races across the sky and sets like a falling rock and at other times it just seems to hang up there all day long.

If Earth were upright in relation to the Sun, still there would be weather, because still there would be cold air near the poles and warm air near the Equator for the atmosphere to contend with. And still there would be the cool and warm variations of night and day. But without the tilt, there would be no seasons.

My people at the Go Figure Academy of Sciences tell me that in a truly upright world, life as you know it would be very different. For one thing, everywhere on Earth all year long would get the same amount of daylight and darkness — exactly 12 hours. For another, there would be no tourist seasons.

As Figure 2-7 shows, this arrangement that gives the Earth the same slant in relation to the Sun throughout the year produces some interesting dates.

✔ On about March 20, the vernal (or spring) equinox, and again on or about September 22, the autumnal (or fall) equinox, it happens that daylight and darkness is distributed evenly around the world — each lasting 12 hours.

✔ Direct sunlight reaches its most northerly point on or about June 21, the summer solstice, over the Tropic of Cancer, an imaginary latitude line 23.5 degrees north of the Equator. In the Northern Hemisphere, this is sometimes called "the longest day" because it is the day of most daylight.

✔ Likewise, on or about December 21, the beam of direct sunlight has reached its most southerly point, over the Tropic of Capricorn, 23.5 degrees south of the Equator. In the Northern Hemisphere, this is "the shortest day," the day of least daylight.

Figure 2-7:
Here is a close-up view of Earth's 23.5 degree slant and how it affects the distribution of sunlight in the course of its yearlong revolution around the Sun.

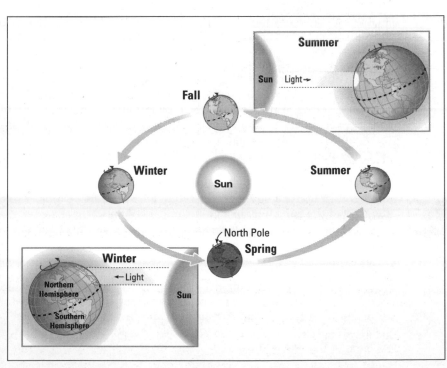

These dates are often said to be the "first official days" of the various seasons, but take another look. Is December 21 *really* the first day of winter where you live? As my people say at the Go Figure Academy of Sciences, this is the kind of thing that can happen when you send an astronomer to do a meteorologist's job! In most parts of the United States, if you haven't been wearing your winter coat before December 21, you haven't been keeping your promises to your mother. Likewise, by June 21, the day of the year when the Sun's rays beat down directly over more of the Northern Hemisphere than any other, summertime has already become a pretty familiar feeling.

In the middle latitudes of the Northern Hemisphere, weather scientists generally think of the winter season as the months of December, January, and February, spring as March, April, and May, and so on. (From winter through autumn, Part III goes into the details of the weather effects of these seasonal changes.)

Spin of the day

You and I are part and parcel of Earth's motions in space, and the planet's atmosphere also is going along for the ride. Although earthlings are traveling in a yearlong revolution around the Sun, speeding through space, they have no sensation of this movement. Thank goodness for that! Talk about being blown away!

Also, people have no sensation of dizziness even though the planet is spinning like a top in a rate of rotation that completes itself every 24 hours. A person standing at the Equator not only is traveling through space along with the planet, but at the same time is spinning with the planet at more than 1,000 miles per hour. It's like a whirligig carnival ride, revolving on one level and spinning on another, and I'm getting a little queasy just thinking about it.

The rotation or spin of Earth has a big impact on daily weather. Every moment, 24 hours a day, a new patch of Earth is being exposed to the warming rays of the Sun after the cooling effects of darkness. Exactly around the world, a patch in daylight is becoming shrouded in the shade of the spinning planet. This constant routine is sending radiant heating and cooling through the atmosphere like a wave.

Such is the fickle pace of many daily weather events. The hot summer afternoon can conspire with the moist air to produce a violent local storm of lightning and thunder and hail. The ground and the town below it can be left in a mess in an hour or two, and before long, the sky can show not the slightest sign of what happened.

Earth's rotation is responsible for some very large and powerful weather-related motions in the atmosphere. For example, weather patterns in the middle latitudes move from west to east because of Earth's spin. It is responsible for the west-to-east direction of the powerful polar jet streams and for the prevailing global winds, such as the tradewinds and the mid-latitude westerlies, which Chapter 4 describes in detail.

Putting on Airs

The atmosphere, the weather's home, begins at the tip of your toes and extends some 80 miles up, more or less. That may sound like a pretty deep sky, but relative to the size of the earth, it's thickness is less than a rind on an orange. And the layer of atmosphere where all the weather takes place is much thinner still — only about 10 miles thick — more like the skin on a peach. Only this skimpy layer, this peach fuzz, contains enough of the ingredients in the right proportion that you and I need to breathe. Does it strike you as odd, by the way, that people so seldom give this vital substance much thought? Go figure. So what exactly is this atmosphere — this precious stuff called air?

Do I smell gas?

The atmosphere is a thin envelope of gases surrounding Earth. Figure 2-8 lays out its contents, within about 50 miles of the surface, and quickly you can see that mostly it is nitrogen and oxygen. No, you can't really smell the gases that make up the air. This mixture is odorless and tasteless. And you can only see it when it contains water drops or something else — when it's dirty. *Then* you smell it and you can see it, even if you don't want to. (Chapter 13 has the low-down on air pollution.)

Figure 2-8:
The gases that make up the atmosphere. The proportion of the gases in the left column are stable, while those in the right column are variable.

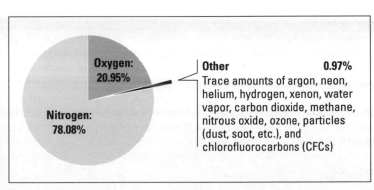

Oxygen: 20.95%

Nitrogen: 78.08%

Other   0.97%
Trace amounts of argon, neon, helium, hydrogen, xenon, water vapor, carbon dioxide, methane, nitrous oxide, ozone, particles (dust, soot, etc.), and chlorofluorocarbons (CFCs)

While it makes up some 78 percent of the atmosphere, nitrogen is not in the weather-making business. It makes the natural nitrogen that is absorbed by the soil and is essential for the growth of plants. It does combine to form nitrous oxide, an important ingredient in smog, which is described in Chapter 13.

The following sections describe the gases that are most important to weather.

Oxygen

Oxygen, the gas that sustains all life on Earth, is constantly being recycled between the atmosphere and the biological process of plants and animals. It combines with hydrogen to form water, which in its gaseous state, water vapor, is the most important component of the atmosphere as far weather is concerned.

The form of oxygen known as the gas *ozone* also is vital to life on Earth. Ozone forms a thin layer in the upper stratosphere that filters out harmful ultraviolet radiation. Chapter 13 describes what happens when this crucial layer is depleted — when an *ozone hole* forms — and what the world's governments are doing about it.

Water vapor

If there is one substance in the atmosphere more involved with weather than any other, it is the gaseous form of water. At its most concentrated, water vapor makes up only 4 percent of the atmosphere, and yet, almost no important weather takes place without it.

Without water vapor to condense into droplets of water or ice crystals as air rises and cools, no clouds would form in the sky.

Without water vapor, there would be no precipitation — no rain and no snow. The cycling of water through the environment, as described in Chapter 3, would come to a screeching halt without water vapor in the atmosphere.

The condensation of water vapor leads to the release of latent heat, which is described earlier in this chapter in the sidebar "How to cause a storm." Latent heat supplies the atmosphere with the energy that is important in the formation of storms, especially thunderstorms and hurricanes.

Water vapor also is a potent gas in the *greenhouse effect,* which is outlined in more detail in Chapter 13. Like the glass top of a greenhouse, it absorbs infrared heat emissions from Earth's surface, preventing it from radiating back into space.

Carbon dioxide

Like water vapor, the gas carbon dioxide has powerful greenhouse effects, trapping outgoing heat radiation, a process that is natural and beneficial — up to a point.

Carbon dioxide is constantly recycled through biological process of animals and plants. It is absorbed from the atmosphere during *photosynthesis,* the process by which plants take sunlight, water, and carbon dioxide and make food to grow on. In the bargain, the plants give off oxygen.

Carbon dioxide is supplied to the atmosphere through such things as the decay of plant and animal material and such human activities as timber harvesting and the emission of exhausts in the burning of *fossil fuels,* such as gasoline and other oil products and coal.

This gas has no direct effect on daily weather, and most of it gets washed out by rain or snow. But like all greenhouse gases, carbon dioxide may have an effect on *climate,* weather's long-term patterns. The concentration of carbon dioxide in the atmosphere has been steadily on the rise for more than a century. (Chapter 12 describes how increased concentrations of CO_2 are leading to concerns about *global warming.*)

The bit particles

The variety of tiny particles of solids and liquids collectively called *aerosols* are not, strictly speaking, atmospheric gases. But these natural and man-made impurities in the air are very much a part of the atmosphere. They play important roles in both daily weather and in longer term climate variations.

On the surfaces of these tiny floating bits of stuff, water vapor condenses to form clouds. In fact, my people at the Go Figure Academy of Sciences tell me that you cannot get a cloud to form naturally without this microscopically small flotsam and jetsam up there. Without clouds, of course, there would be no precipitation, and without precipitation — well, you get the picture. There would be no *Weather For Dummies!*

The atmosphere is carrying tons of aerosols, such things as soot and ash from fires, dust kicked up by winds, sea spray, and large quantities of ash and droplets of gases from the eruption of volcanoes. In fact, the gigantic eruption in 1991 of Mt. Pinatubo in the Philippines threw so much material high into the atmosphere that it changed the short-term global climate. Aerosols from this single volcano filtered out sunlight, causing average surface temperatures in the Northern Hemisphere to decrease by nearly 2 degrees. This worldwide cooling effect lasted nearly two years.

On the downside, industrial processes and gasoline engine combustion releases enormous quantities of aerosols, which is responsible for the formation of smog in urban areas. Chapter 13 describes these and other air polluting effects of aerosols, including acid rain.

How high the sky?

There is no well-defined upper boundary of the atmosphere (see Figure 2-9). It just gets thinner and thinner and thinner. It's like the thin skin of a peach — clearly defined on one side where it covers the fruit and then fuzzy on the other.

You could say that the atmosphere extends roughly 80 miles up, because a few molecules of the lighter gases like helium and hydrogen are still drifting around up there. But 99 percent of Earth's sky stuff is below the top of the *stratosphere,* 30 miles in the sky, and 80 percent of it is in the lower layer, the 10-mile-thick *troposphere.* And as far as weather is concerned, and as far as breathing is concerned, you and I are out of business at the top of the *troposphere.*

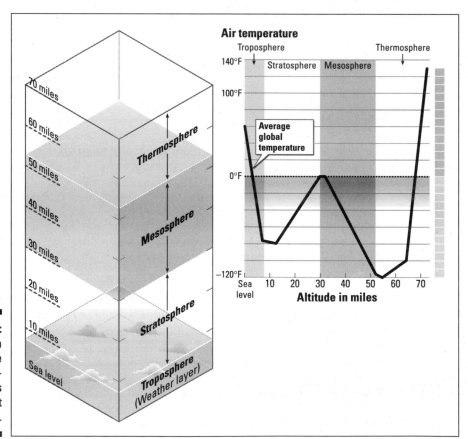

Figure 2-9:
Here is a map of the atmosphere's different layers.

Chapter 3

Land, Sea, and Precipitation: Is This Any Way to Run a Planet?

• •

In This Chapter

▶ Following the water cycle from the sea — and back

▶ Sorting out the forms of precipitation and condensation

▶ Riding air's roller coaster across the landscape

▶ Going with the flow of the big ocean currents

• •

*E*arth's surface is mostly water. It's easy to think of the planet's land masses as the places where all of the important business is done. Being land creatures, of course, you and I have a certain bias in this respect. But when it comes to making weather, the water is where a lot of the action is.

Living in cities, it's easy to get into the habit of thinking about the weather only as a series of inconveniences. Have you noticed how often airports and highways are the scenes of television news reports about storms? Travelers often are the first to feel their effects. But rain and snow do more than make travel difficult or slow-going for you and me.

In this chapter, the role of water gets its dew, so to speak, and all forms of precipitation get a closer look. Here, you can find the answer to the question, "What is fog, exactly?" And here is a description of how the oceans and the land masses affect your weather.

Water's Stirring Role

Precipitation, the liquid and solid water particles that fall from the sky, is part of a large and truly amazing cycle of water. It passes from the bottoms of the seas to the tops of the mountains, from the heights of the sky to the depths of the earth. Changing water's form from gas to liquid to solid, weather drives this water cycle that is the heart of the possibility of life.

Notice this: When scientists look for signs of life on other planets, one of the first things they look for is a sign of water. Sure, they're inconvenient, all of these storms. Hey, no planet is perfect!

The atmosphere takes up water from out of the warm, salty ocean and delivers it cool and fresh to the land through precipitation of storms. The chilling cold of winter stores it in the mountains as snow and ice. The warming spring brings some of it down the rivers and into the lakes. Over the following days and months and many years, the fresh water flows back to the salty sea. One way or another, all of the water you drink — and everything else you drink — and all of the food you eat depends on the storms, recent storms or storms of the distant past. Figure 3-1 illustrates the basics of the water cycle.

The water cycle is constantly transferring the world's water supply between these basic storage reservoirs: the oceans, the land, and the atmosphere.

The salty oceans contain 97 percent of Earth's water. This means that on any given day, only 3 percent of the water in the world is fresh water.

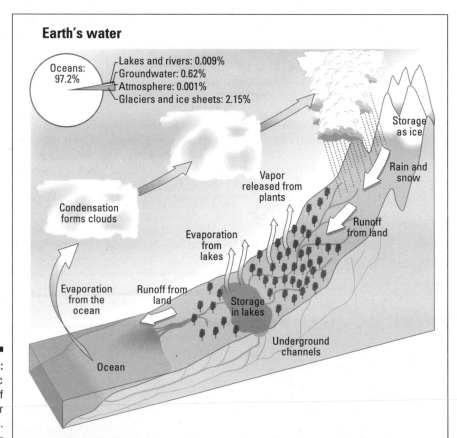

Figure 3-1:
The basic elements of the water cycle.

A world of water

Viewed from space, the colors of Earth, the best looking planet of the lot, are mostly blue and white. It is a world of oceans and clouds, for the most part, its poles capped by snow and ice.

Most maps of the world don't give you a very good sense of this, but 75 percent of Earth's surface is covered by oceans, lakes, streams, and ice sheets. Projections that focus on the Northern Hemisphere — where land masses cover 39 percent — have a way of abbreviating the vastness of the Pacific Ocean, a region that is hugely important to world weather.

If following weather systems is what you're interested in, find yourself a map that doesn't cut the Pacific Ocean in half. Better yet, find yourself a globe and get the real picture.

Here's where that skimpy 3 percent of the world's water that is fresh is to be found:

- A whopping 75 percent of it is locked away in glaciers and ice sheets.

- Most of the rest — 24 percent — is what is known as *groundwater,* stored as soil moisture, as seepage in rocks, or in underground reservoirs called *aquifers.*

- Rivers and lakes hold only .33 percent — a third of one percent — of all freshwater.

- The atmosphere contains only .035 percent of all fresh water.

My people at the Go Figure Academy of Sciences have done the math. Amazingly, all of the water in all of the lakes and rivers, all of it down in the ground and up in the atmosphere — the whole shebang amounts to only about 1 percent of the water in the world. This 1 percent supplies all of the rain and all of the snow that the weather throws at you and me and meets all of the daily needs of life.

Driven by the heat energy from the Sun, water is cycled between its three phases or physical states — liquid, gas, and solid — in its journey between the oceans and other bodies of water, the atmosphere, and the land. The cycle is a continuous loop, of course, although some parts of it occur in only a matter of minutes, while other phases of the process drag on for thousands of years.

The following sections outline the major steps in the water cycle.

Ocean to atmosphere

Water leaves the ocean and becomes part of the atmosphere by being transformed from its liquid state to its gaseous form, known as *water vapor,* and leaves the briny salt in the sea. This happens through the process of *evaporation,* which I describe later in this chapter's section "Rain to Rime: Forms of Precipitation." This change of physical form absorbs heat, storing away a bit of the Sun's energy in the water vapor molecule as it rises up into the atmosphere (see Chapter 2). Not all of the briny salt is left in the sea. Some of the salt breaks out into the atmosphere, where some is carried aloft and eventually serves as the condensation nucleii described later in this chapter.

Water is evaporating from liquid into water vapor everywhere across the surface of the oceans — and across most continents, for that matter — all of the time. But it happens especially during the warmer times of day, and especially at the warmer places on Earth.

Weather scientists estimate that of all the water that enters the atmosphere as water vapor, 84 percent of it evaporates from the ocean, and 16 percent from the continents. So how much water evaporates from the ocean every year? About 92 quadrillion gallons sounds about right. That's 92,000,000,000,000,000.

The region that contributes most of this water vapor to the atmosphere is the Tropics, the warm band around the Equator that receives most of the direct sunlight. In the Tropics, huge volumes of water vapor evaporate into the atmosphere. Bands of giant thunderstorms take shape north and south of the Equator. These storms irrigate the tropical rainforests of Asia and Africa. The clouds in the Tropics also form in big bands of storms over the warmest water near the Equator in the Pacific and Atlantic oceans where tradewinds converge. The five-dollar word for this area is the *Intertropical Convergence Zone* (ITCZ).

Atmosphere to surface

A molecule of water vapor evaporated into the air will spend only about a week, or maybe ten days, in the atmosphere before it condenses and falls back to the surface as one form of precipitation or another. Its life in the atmosphere may be only a matter of hours if it becomes part of a thundercloud that develops rain or hail. Or it may become part of warm light rain called *virga* that evaporates back into the atmosphere even before it reaches the surface. It may fall as rain on a sizzling hot parking lot and return to the atmosphere in a flash. It may fall as snow on the top of a giant mountain and be there 1,000 years.

A molecule of warm water vapor can rise miles into the atmosphere and travel far across the world. In fact, scientists say, it usually does. They have drawn a picture of how far a typical water vapor molecule travels from the point where

It's just a phase

Only the atmosphere contains all three physical states or phases of water: gas, liquid, and solid. As a gas, it is present as water vapor. As a liquid, it is present as tiny droplets in clouds and as falling rain. As a solid, it is present as ice crystals in clouds, as sleet, snow, and hailstones.

As heat energy and pressure is added or subtracted, water changes from one form or another in a process called a *phase change.*

Water changes from one phase to another by these processes:

✔ *Evaporation* liberates water molecules from the liquid state into its gas form, known as water vapor.

✔ *Condensation* causes water vapor to connect to other water molecules and form a liquid.

✔ *Freezing* causes liquid water molecules to form a solid.

✔ *Melting* causes water to move from its solid to its liquid phase.

✔ *Sublimation* allows water to transform from its solid state of snow or ice directly to gaseous water vapor without first becoming liquid. Sublimation occurs in clouds, when water vapor forms ice crystals

it evaporated into the air. Going with the flow of the atmosphere's winds, according to one study, it is likely to be carried 6,000 miles east or west and 600 miles north or south.

So how much water vapor is up there? By some estimates, the atmosphere at any given time is holding up to 25 million billion pounds of water. If it all condensed and fell at once, the Earth would be covered by an inch of water. A lot of cycling is going on. In a year, the atmosphere produces an amount of precipitation around the world that is more than 30 times its total capacity to hold water.

The sections in this chapter about precipitation and condensation describe the variety of ways that the atmosphere has of delivering this water vapor back to the surface and what it takes to make it happen.

Surface to ocean

For much of the water vapor in the atmosphere, the journey back to the ocean is very direct. Most precipitation falls as rain, and most rain falls on the oceans. Weather scientists estimate that 77 percent of the world's precipitation falls in the ocean. It stands to reason when you think about it: Oceans cover 71 percent of the planet's surface, contain 97 percent of the water, and contribute, through evaporation, 84 percent of the water vapor to the atmosphere. And all of those storms in the Tropics produce an awful lot of rainfall. If you look on a map at the Tropics — that region 23.5 degrees north and 23.5 degrees south of the Equator — you'll see that most of it is ocean.

Setting the water table

The *water table* is the boundary between the layer of soil that has both air and water in its gaps and the layer that is saturated with water. Add water to the ground, and the water table will rise.

If you're digging a well, the water table is the level underground that you want to get below — and stay below — so that your well has water in it all year long.

Because it absorbs water, the layer above the water table plays an important role during times of heavy rains or spring runoff. When this layer becomes saturated or if it is frozen, the area becomes especially likely to flood because the new water has nowhere to go but to flow over the surface.

For all your living needs, you depend on the other part — on the other 23 percent of the world's precipitation that *doesn't* fall immediately back into the ocean. Your life and my life depends on the water vapor that condenses into precipitation and falls on land, and on the time in the water cycle that it spends as fresh water before returning to the sea. In an average year, the weather delivers 30 inches of water that falls unevenly across the United States. From this 30-inch layer, just under 18 inches returns to the atmosphere by evaporation. About one-half inch seeps into the ground. And the remaining amount, not quite 12 inches, is the surface water that is flowing downhill in streams and rivers and stops for a while in lakes before returning to the nearest ocean.

Rain to Rime: Forms of Precipitation

Moist air cools as it rises, and when it reaches a certain point the water vapor in it condenses into water droplets. Warm air can hold more water vapor than cool air, so anytime water is cooled it is moving toward saturation. This point of saturation is called the *dewpoint*. As air continues rising and cooling, eventually it reaches saturation. The droplets that the water vapor forms are not raindrops, but much smaller — typically about one thousandth of an inch across. These droplets are the stuff of clouds and fog. It takes about a million microscopic droplets to form an average raindrop. And something else has to be up there in that cloud for precipitation to form.

Weather scientists still do not understand all of the circumstances necessary for clouds to begin dropping rain. Still there are mysteries in the clouds. But they know that the droplets of water vapor need a place to meet in order to form raindrops or ice crystals in the clouds. They need the surface of microscopic particles of dust or other material called *aerosols*. (Chapter 2 describes these aerosols.) If the cloud is cold enough, ice crystals will condense around these minute particles, and the crystals will join together to form snowflakes. In a relatively warm cloud, the droplets condense around a particle and will grow larger by bumping into one another until raindrops are formed. (Chapter 5 details the formation and characteristics of the different kinds of clouds.)

Rain and snow account for all of the big, important amounts of water that falls from the clouds around the world, but the atmosphere deposits precipitation in a variety of liquid and frozen forms.

Rain

By far the most common form of precipitation, *rain* is a drop of liquid water that has become too heavy to remain in its cloud. In many parts of the world, rain is really the *only* form of precipitation. (Chapter 9 goes into more detail about rain and rainstorms.)

More often than not, the rain that falls in your face is melted ice. This may not be true in the Tropics, however, where the rain is water that has remained in its liquid state from the top to the bottom of the cloud.

Getting the dewpoint

Two everyday ways are used to describe the moisture in the air. One is *relative humidity,* and the other is *dewpoint.*

Relative humidity is the amount of water vapor in the air relative to its saturation point. At 50 percent humidity, air is half saturated. It sounds simple enough, but it has this one flaw as a handy-dandy comfort indicator: The saturation point changes with temperature, because more water vapor can exist in warm air than in cold air. Up goes the air's temperature, down goes its relative humidity, even though it contains the same amount of water vapor.

As a comfort indicator, dewpoint has a clear advantage over relative humidity, although it takes a little getting used to because it's expressed as a temperature. It doesn't tell you directly about water vapor in the air, but it tells you more directly what you want to know.

Dewpoint tells you the temperature of the air at saturation. The advantage is that dewpoint just as accurately reflects the water vapor in the air, but it doesn't change with temperature.

When the dewpoint temperature is high, the moisture content of the air is high, but this indicator won't be moving all day like relative humidity — not with this air, anyway.

The key is the distance between the air's temperature and its dewpoint temperature. If the distance is big, the air is dry, and chances are it won't come near saturation and condensation will not occur. If the distance is small, the moisture content of the air is high, and you could be in for an uncomfortable day and night.

As a rule, dewpoints in the mid-50s make comfortable summer afternoons, but everybody's comfort zone is different.

Even in summer, the storms of the middle latitudes often produce rain that began as an ice crystal and then formed into a snowflake that melted and collapsed into a raindrop as it fell through warmer layers of air.

Raindrops range in size from about two-hundredths of an inch, 0.02 of an inch, to roughly a quarter of an inch, at which point they begin breaking up into smaller drops. A typical raindrop is about a 16th of an inch. That fine drifty stuff that falls from the sky in drops of less than .02 of an inch is called *drizzle*.

Is a raindrop shaped like a teardrop? No, at no time is a raindrop shaped like a teardrop. (Come to think of it, is a tear?) If it is a relatively small drop, it is round as it falls. A larger drop shows the effects of air resistance building underneath it. Its bottom flattens and its sides bulge as it falls.

Freezing rain is supercooled raindrops that freeze on impact with cold surfaces. Storms that produce freezing rain, called *ice storms,* are some of the most dangerous and damaging of all weather events. The weight of the accumulating ice crushes trees and snaps electrical power lines. Layered by nearly invisible ice, roadways become extremely hazardous. Whole herds of livestock and flocks of birds can be wiped out. (Chapter 8 describes ice storms and freezing rain in more detail.)

An advancing warm front can lead to ice storm conditions as rain produced by the elevated warm air mass falls through colder air near the surface. (Chapter 2 describes warm fronts.)

Snow

Snowflakes are collections of ice crystals that assemble themselves as they fall through a cloud. Much precipitation forms first as ice crystals and then snowflakes in the winter storm clouds of the middle latitudes, although often it melts into raindrops as it falls into layers of warmer air.

Snowflakes come in different shapes and sizes and moisture contents. As every skier knows, there are big, soggy snowflakes that sometimes melt into clumps as they fall quickly through warm air, and there are small, light, and billowy flakes that form great dry powder. If it's skiing you're up for, hope for the dry powder. If it's water you're interested in, it's the wet snow you want. It takes about 5 inches of wet snow to yield an inch of water, 10 inches of average snow melts down to 1 inch of water, and 20 inches of dry snow — and occasionally as much as 50 inches — to give an inch of water. (There's more about snow and snowstorms in Chapter 8.)

Too cold to snow?

Everybody who spends a winter in places where the snow falls has heard the expression "It's too cold to snow." The thinking is that snow falls within a certain range of temperatures, and when the air gets colder than that, when it gets *real cold,* it's too cold to snow.

My people at the Go Figure Academy of Sciences have looked into this, and lo and behold — no. It is never too cold to snow. Cold air does not contain as much water vapor as warm air, it is true, but even very cold air contains enough to make snow.

It happens, of course, that the coldest days of winter often are days when there are no clouds in the sky. It is certainly true on these days: It's too clear to snow.

Hail

Hail is a large frozen raindrop that is formed inside the enormous cloud of an intense thunderstorm. The cloud contains a powerful updraft that keeps the frozen raindrop from falling. Supercooled droplets freeze onto the ice, forming frozen layers. Eventually, the hailstone is too heavy for the updraft to support any longer, and it falls to the ground.

Hailstones range in size from two-tenths of an inch, .20 of an inch, to really big ones that can grow to the size of a softball or even larger.

Have you ever noticed how sometimes hailstones have that wet glazed look of melting ice? They have just fallen through the warm, rainy bottom of the giant thundercloud before bouncing onto the ground. The bouncing is a telltale sign of hail. (Chapter 10 describes hail and thunderstorms in more detail.)

Graupel (snow pellets)

Sometimes ice crystals fall through a cloud of supercooled droplets — minute cloud droplets that have fallen below freezing temperature but have not frozen. The ice crystal plows into the supercooled droplets, and they immediately freeze to it. This process forms *graupel,* or snow pellets, as the droplets continue to accumulate on the crystal. The pellets bounce when they hit the ground.

Sleet (ice pellets)

Out ahead of the passage of a warm front, falling snow may partially melt and then refreeze into a frozen raindrop before it reaches the ground. These ice pellets, easily visible white stuff that bounces off the ground under these conditions, are called *sleet*.

Sleet forms under conditions that are similar to those that produce freezing rain. The difference is that the lower layer of cold air is deeper so that the partially melted snowflake or cold raindrop has time to freeze into an ice pellet before reaching the ground. Sleet can accumulate massive layers of ice, but its effect is different than the ice layers formed by freezing rain. The pellets do not accumulate on trees and power lines like freezing rain, and because it is formed by pellets, the ice layers of sleet are less slippery. But watch your step.

Because it is easily seen and does not accumulate layers of ice, the impacts of sleet are less dangerous than freezing rain.

Rime

Rime is a milky white accumulation of supercooled cloud or fog droplets that freeze when they strike an object with a temperature that is below freezing. The process is called *riming* when supercooled cloud droplets attach to ice crystals in the formation of snow pellets or graupel. (See the section earlier in this chapter for more on graupel.)

When the supercooled fog or cloud droplets accumulate on the ground on trees or telephone poles or the sides of buildings, the bright white stuff is called rime ice or rime. This form of ice is white rather than transparent because the droplets have trapped air between them as they froze.

Rime ice can pose a hazard to an airliner when it forms on a wing as an aircraft flies through a cloud of supercooled droplets. The droplets freeze immediately onto the wing and can form an irregular surface, redistributing the flow of air. But rime is relatively lightweight ice and easily removed by de-icers.

Dew to Fog: Forms of Condensation

Most water vapor in the atmosphere at any given moment is not in the business of meeting in those large conventions of minute water droplets or ice crystals that you and I see as clouds. So you don't see most water vapor. Its molecules just hang out there in the air like the invisible molecules of the atmosphere's other gases. For air to produce something you can see, it has to become saturated with water vapor — it has to reach its *dewpoint*. All things being equal, warm air can carry more water vapor than cold air.

For water vapor to become visible, it requires a process known as *condensation* — its conversion from a gas into a liquid — or *deposition* — its conversion from a gas directly into solid ice. (These processes are outlined in detail in Chapter 5 as they lead to the formation of clouds.) But condensation and deposition also happen closer to home.

Forms of condensation that show up on the ground play only bit parts in the global water cycle. They are quiet and minor players in the atmosphere's stormy pageant of weather. But you and I know their work intimately, and they are not to be trifled with. Often they come in the night, and occasionally they catch travelers unaware of the special dangers they pose.

Dew

During the night, most often on a clear and calm night, when objects on the ground radiate away the Sun's daytime warmth, they can become colder than the shallow layer of air surrounding them. If the ground surface cools to the point where the air just above it is saturated — reaching the temperature known as its dewpoint — water vapor begins to condense on blades of grass and what-have-you. This condensation is *dew*.

Dew can form beads that drip from the leaves of plants and trees or can form a continuous coating of liquid water on surfaces.

Dew can be an important source of water to plants during dry periods, yielding as much as 2 inches of water in some areas over the year.

Frozen dew

Should temperatures of the surface continue to fall below the freezing point, the condensed beads of water become beads of ice, or the layer of water becomes a sheet of ice. *Frozen dew* can pose a real inconvenience as well as a serious driving hazard to motorists.

Sometimes it's frozen dew that forms a layer of ice that sticks so stubbornly to the windshield of a car left out overnight. Frozen dew can make it difficult to work the lock on the car door, and it can freeze the door to the frame. Worse still, frozen dew can form *black ice,* a particularly dangerous patch of frozen roadway because it is so difficult to see. Bridges, which sometimes are colder than the rest of the roadway, can be especially dangerous in conditions that produce frozen dew.

Frost

On a cold, clear, and calm night, the same surface-cooling that would produce dew will instead produce *frost* if the dewpoint temperature — frostpoint temperature — of the air just above it is at or below freezing. When frost forms, the water vapor in the air is converted directly from a gas to solid crystals of ice.

Unlike the beads of frozen dew, frost forms delicate white crystals of ice in treelike branching patterns that decorate the windowpanes of winter. Sometimes frost is called *white frost* or *hoarfrost*.

Fog: A grounded cloud

Fog is a cloud that's grounded, but it doesn't take shape like a cloud. While clouds form by air rising and cooling to its dewpoint, most forms of fog are not the result of rising air. Fog forms near the cold ground that cools the air above it to its dewpoint, much like frost and dew forms. Clear nights favor the formation of fog because the air near the ground gets colder than it does when there are clouds in the sky.

As the air temperature nears its dewpoint, the formation of minute cloud droplets begins as a gradual process, first giving the air a hazy look and then becoming more dense.

For good reasons, dense fog is notorious for its treachery on roadways as well as its ability to close airports. Because fog forms, like other water droplets in clouds, around tiny particles in the air, the dirtier air of cities makes thicker fogs. Polluted air also can produce unhealthy acid fog, which Chapter 13 describes. Even when it is not particularly polluted, a long siege of thick fog can make for a miserable feeling of claustrophobia.

Fog aficionados use terms like *burning off* and *settling in* and *lifting* to describe its comings and goings. Fog burns off or dissipates when the Sun warms the ground, forcing temperature of the nearby air above its dewpoint and evaporation quickly sets in. Sometimes a fog layer is too thick to burn off. Even modest warming of the ground can cause enough daytime radiation to lift fog off the ground. It may hover overhead as a low stratus cloud and then descend again overnight as the radiating ground cools.

Fog comes in a variety of flavors:

✔ *Radiation fog* or *ground fog* forms upward from the ground when air chilled by the cooling ground underneath becomes saturated with water vapor — reaches its dewpoint. Ideal conditions for this fog are a calm night when a shallow layer of moist air is covered by drier air. Because the cold air is heavy and collects in the lowest ground, this is often called *valley fog*.

- *Advection fog* is a fancy name for fog that forms in one place and is blown by winds (or advected) to another. This fog is common to the U.S. West Coast, for example, when sea breezes bring relatively warm and moist Pacific Ocean air over a strip of much colder water along the coast. Fog forms as the cooling air reaches saturation and often the breezes push it inland. This is the stuff that inspired Mark Twain to say, "The coldest winter I ever spent was a summer in San Francisco!" In the eastern U.S., advection fog forms most often when warm moist air is blown over a snowpack. In the southern and central U.S., advection fog forms during winter over land when warm, moist air from Gulf of Mexico is blown in over the cold ground.

- *Upslope fog* forms when moist air is pushed up the slope of a hill or mountain and cools as it rises sufficiently to reach its point of saturation.

- *Steam fog* or *evaporation-mixing fog* are terms used to describe fog that forms over warm bodies of water, most often in the fall. Steam fog forms over a heated swimming pool. Also, this is the stuff your breath makes when the warm, moist air from your lungs hits the cold, dry air of the winter morning. It's the addition of the water vapor that quickly saturates the cold air. You're forming a tiny cloud in front of your face with every breath you take.

Weather and the Land

The atmosphere is very impressionable. What would you expect of something that spends all day blowing from one place to another? The air that makes the weather picks up the slightest changes in temperature and moisture it encounters. It is influenced by the shape and even by the texture of the surface it flows over. Whether it is flowing over water or land, for example, makes a big difference.

A lopsided planet

Weather has a lot to do with the atmosphere contending with forces that are unevenly distributed around the world. The planet does a masterful job at creating these conditions. As Chapter 2 points out, its midsection gets much more of the Sun's energy than anywhere else, for example, and its tilt makes for strikingly different seasons in the year.

Earth has another interesting irregularity. The continents are not evenly distributed over the globe — not by a longshot. The land masses are disproportionately located in the Northern Hemisphere, north of the Equator.

Take a look at Figure 3-2, the world from directly over the North Pole. Land covers 39 percent of its area from here, and the ocean accounts for 61 percent. Now check out the flip side, from directly above the South Pole. This half is 19 percent land and 81 percent water. Differences like these make for a lot of lifting and hauling of energy around the world — and a lot of weather.

Radiating hot and cold

O, give me land, lots of land, and I'll give you lots of weather. This is where the big contrasts in temperatures abound. And sharply contrasting temperatures help make a lot of weather.

Land quickly heats up during the day and quickly cools off at night, changing the weather above it almost hourly under certain conditions. In the longer term, deserts and snowfields send very different signals to the air flowing over them.

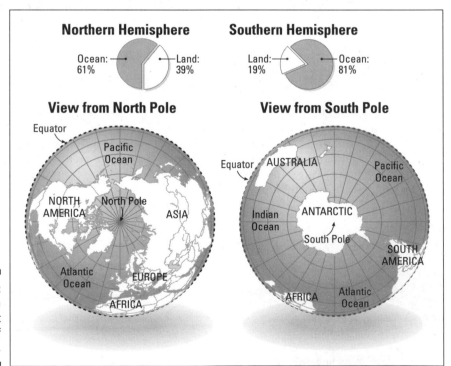

Figure 3-2: The different faces of Earth.

Air's roller-coaster ride

The shape of the landscape is a powerful maker of clouds and storms. Winds force the air up one side of a hill or mountain and down the other side. On the way up, its temperature cools and its volume expands, and clouds and storms often result. And on the way back down, the air warms and contracts, and the clouds evaporate.

The Rockies, the Sierra Nevada, and the Cascades are examples of mountain ranges that produce these *orographic* clouds and storms. Figure 3-3 illustrates how these work. In western Washington, there are places on the western side of the Cascade mountains where precipitation totals average as much as 180 inches per year. In eastern Washington, in the "rain shadow" on the eastern side of the mountains, precipitation averages less than 10 inches per year.

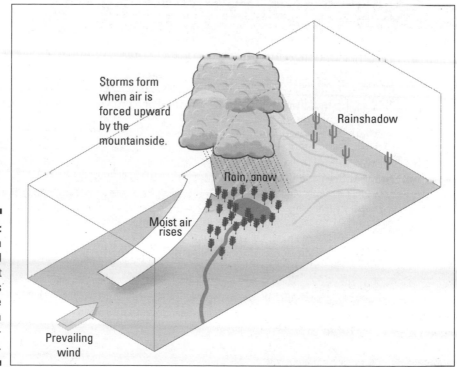

Figure 3-3:
This is a typical pattern that develops as winds force air up a mountain-side.

Weather and the Ocean

Without the oceans, the weather would be just a lot of hot air. Well, that's an exaggeration, but you get the idea. Land surfaces make good storms, but most of the rain and snow that falls is water vapor that has evaporated from the ocean. As this chapter's section on the water cycle points out, that's where almost all that water has come from, and that's where it's going.

Up and down the U.S. West Coast, virtually all of the storms of winter are dumping Pacific Ocean water vapor. Even in the nation's midsection, the water vapor that becomes most rainfall is an import. In the Mississippi River basin, according to one study, as little as 24 percent of precipitation comes from local evaporation.

In coastal areas, the boundary between the ocean and the land make interesting local weather conditions. Sharp contrasts between slowly changing sea surface temperatures and rapidly changing land temperatures can sometimes cause dramatic shifts in winds and temperatures during the day.

On a bigger scale, sea surface temperatures in the tropical Pacific seem to cause major shifts in the jet stream and steer storms over North America. Weather scientists are still figuring out how big a role Pacific Ocean conditions play in determining global weather patterns that make one winter different from another. (Chapter 6 describes El Niño and similar ocean conditions that affect weather across the U.S.)

Warm ocean currents such as the Gulf Stream carry warm air that gives the U.S. southeastern coast a much more tropical feel than it otherwise would have. The same current continues across the North Atlantic and bathes Europe in a warmer climate than its high latitudes would suggest. In contrast, on the U.S. West Coast, the California Current carries cooler water and cooler air temperatures southward. Figure 3-4 is a good look at some of the major weather-altering ocean currents around the world.

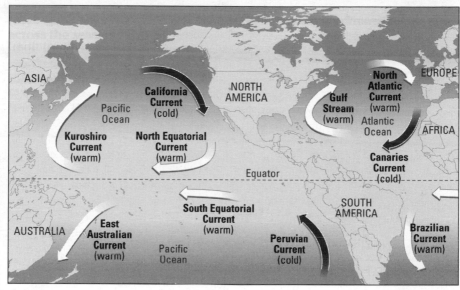

Figure 3-4:
The
California
Current and
Gulf Stream
affect U.S.
weather.

Part II
Braving the Elements

The 5th Wave   By Rich Tennant

"Get in the cellar, Ma—twisters heading this way, and it's a big one!"

In this part . . .

Looking for answers to some basic questions about the weather? You've come to the right place.

Want to know what makes the wind blow? Want to know the names for all of those clouds? Want to know the difference between weather and climate?

In this part are the answers to all these questions and more like them. And you might be surprised. For a lot of people, when they start discovering the names for the cloud types, they start noticing them more. Depending on where you live, of course, it's a skill you can practice 'most every day just by looking up into the sky. It's fun. It's one of the rewards for the work in this part: getting up close and personal with the basic elements of weather and climate.

Chapter 4

Blowing in the Winds

● ●

In This Chapter

▶ Flowing with the highs to the lows

▶ Flying the jet streams

▶ Sailing the tradewinds

▶ Breezing through monsoons, Chinooks, and Haboobs

● ●

*O*ne of the most powerful forces of nature is something you can easily sense but can't really see. You can see signs of it, of course, and its effects are very well known. You can feel it on your face as a stab of cold or a gust of cooling relief. On television every day, you can see satellite images that show swirling circulations of clouds that it carries across the face of the globe like a giant invisible hand. Outside your window, you can see the leaves it stirs and the dust it kicks up. You can hear it flapping a flag or banging shut an open door. This force of flowing air can be delicate enough to give a butterfly flight and on the same day be so brutal it disintegrates a house and everything in it. It is much feared and for very good reason, because people die violently in the commotion it creates. So what goes on up there? Why all of this wind?

Like most everything else in weather, the wind is tied ultimately to the heat energy from the Sun, and the directions it takes are shaped by the rotation of the planet. This chapter explores the reasons for the winds. It describes the tradewinds and westerlies, patterns of the atmosphere that the world's sailors have relied on for centuries. It discusses the high-flying jet streams that influence the storms that cross the face of the planet. And it identifies the seasonal regional winds such as the monsoons, the Santa Anas, and Chinooks.

Taking the Pressure

Wind begins with *air pressure,* a characteristic of the atmosphere that sounds like so much gobbledygook to a lot of people who are simply trying to decide what clothes to wear the next day. But air pressure is fundamental to the workings of winds. Air pressure goes up and down in response to the shifts in the weight of the atmosphere. These shifts are the driving forces of winds. The air is incredibly sensitive to these movements — these shifts in *pressure differences* — and is constantly rushing around trying to even them out.

Forecasters talk about air pressure from time to time, but it's the part of the weather news that a lot of people are likely to sort of tune out. You know, it's the yadayadayada. Hey, nobody plans their day around air pressure, right? You don't need an umbrella or a big hat to protect against it. Anyway, what's with all of these *ridges* and *troughs?* Why all of these *depressions?* And exactly where is this place *aloft?* In this part of the forecast, the meteorologist explains *why* she predicted the stuff you really want to know. And this is when she is likely to give herself a little wiggle room, by the way, while you're not listening. She's covering, well, you know, her bases.

Hey, listen up. If it's weather you want, you're going to have to get some air moving. If you want to move air, or anything else, for that matter, you're going to have to exert some force — some pressure.

Pains and popping ears

Some people are so sensitive to air pressure changes that they can feel the subtle differences between high pressure and low pressure across the surface of the landscape. They swear they can feel in their bones when a storm is coming. Whether they really can or not, consider this advice: Don't argue with somebody in pain.

Conditions such as arthritis or the lingering effects of an old injury are likely causes for such sensitivities. Possibly, tiny pockets of air in their joints bring pressure on their nerve endings when small changes in surface air pressure arrive.

For most people, thankfully, it takes the more dramatic changes in *vertical* air pressure to feel the discomfort of its effects. Changing pressure is what makes your ears "pop" after takeoffs or landings in an airplane or a drive up or down a mountain or even in a fast-moving elevator in a tall building. There's a tube that runs from your middle ear to the top of your throat, and the pop you hear is the air suddenly passing through it, balancing the pressure in there.

The cabins of airliners are pressurized to about 75 percent of normal sea level air pressure. This level of pressure keeps everybody comfortable cruising up there above 18,000 feet where air pressure outside is less than half of normal and the supply of oxygen is so small you would get dopey pretty fast.

Visitors to high mountains, above 8,000 feet elevation, can encounter a condition known as *acute mountain sickness* if they stay more than a day or so and don't take precautions. The body is working especially hard up in the thin air and needs special care.

Differences in pressure is one of the things that causes air to move up and down. As Chapter 5 describes in more detail, when air moves from one place to another across the countryside, other air has to move out of the way. When air is circulating inward from all directions toward lower pressure, it forces other air to rise in the sky. When air is moving up, often you've got cloudiness and storms to contend with. Air circulating outward from high pressure has the opposite effect of pulling air toward the surface. When air is moving down, often skies are clear. (Chapter 2 describes storms, and Chapter 5 is all about clouds.) And differences in pressure make air move sideways. Winds in the upper atmosphere make the air masses march into one another. At the surface, it makes what blows in your face.

Think of the atmosphere as a blanket of gas that is held to Earth, like everything else, by gravity. The mass of the planet pulls at the mass of the gas molecules in the air, the same way it pulls at the molecules in an apple falling from a tree. This clinging blanket of air is never quite evenly laid out or entirely at rest. (What's going on under there?) The atmosphere's weight is constantly shifting. It's always being heated by energy from the Sun in some places and cooled in others. And the force exerted by Earth's rotation is always twisting things, making wrinkles.

A World of Wind and Pressure

A pattern of general circulation of the atmosphere (technically called the *General Circulation of the Atmosphere*) helps describe the fairly constant presence of the big winds and pressure systems around the world. This pattern accounts for the big belts of prevailing winds that seem always to be blowing. These are the westward flowing *tradewinds* near the Equator, the eastward flowing *westerlies* of the middle latitudes, and the *easterlies* of the polar regions.

Like most things in weather, the general circulation pattern of the global winds is driven by energy from the Sun. They are part of the constant motions of the atmosphere as it seeks to balance the heat in the air along the Equator with the cold in the air at the poles. Figure 4-1 is a diagram that weather scientists use to explain the worldwide pressure patterns and the wind systems they create.

Its most dominant feature is the warm air rising high into the atmosphere above the Tropics. This air circulates toward the poles in a cell that sinks in a band of high pressure at between 20 degrees and 30 degrees north and south. In the Northern Hemisphere, this sinking air keeps storms away and helps explain the desert climates of northern Africa and the southwestern United States.

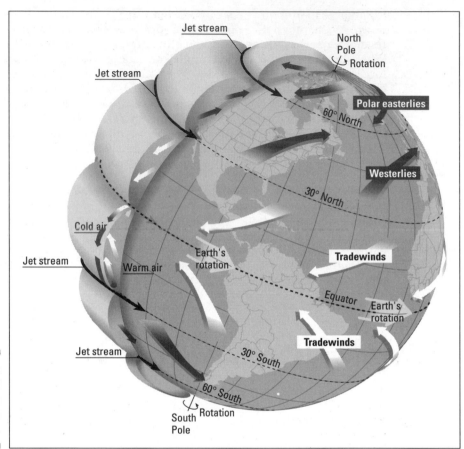

Figure 4-1:
The general
circulation
of the
atmosphere.

Air spreading from this subtropical high pressure partly returns toward the Equator. This return flow near the surface fuels the tradewinds. These persistent surface winds are circulating clockwise around high pressure in the Northern Hemisphere and counterclockwise around high pressure in the Southern Hemisphere. This arrangement causes big east-to-west flows in regions north and south of the Equator.

Air descending from this big high pressure belt also circulates farther toward the poles, away from the subtropics. This flow fuels the westerlies, which drive weather systems from west to east across the heavily populated regions in the middle latitudes of the Northern Hemisphere.

Much of this region lies in what is known as the *Subpolar Low*. In this band, areas of warm air moving north from the subtropical high pressure and cold air moving south from high pressure near the poles come together and often do battle.

Weather scientists are quick to point out that the real circulation patterns of the atmosphere are a lot more complicated than this scheme. The pattern in the Northern Hemisphere is especially complicated by other forces responding to the distribution of ocean areas and land masses. Still, the general circulation of the atmosphere explains a lot. For example, it helps explain the presence of some big persistent high pressure systems in the Atlantic and Pacific oceans that shape a lot of summer weather in the United States. And it helps explain the big low pressure systems that linger through the winter off the western and eastern coasts of Canada that shape so much of that season's weather across the U.S. and Canada.

That muggy Bermuda High

Do you know why the eastern two-thirds of the United States has summers that are so often humid? The general circulation of the atmosphere helps explain it. At about 30 degrees north latitude, in that band of sinking air formed by the cell of circulation from the Equator, a high pressure system squats out in the North Atlantic Ocean.

As it does around all high pressure areas in the Northern Hemisphere, air circulates clockwise around this high pressure system, known as the North Atlantic High or the Bermuda High. As Figure 4-2 shows you, the winds circling south and west of this high pressure carry into the eastern United States a persistent flow of muggy, subtropical moisture. Chapter 1 explains why humidity makes high temperatures more uncomfortable. The moisture flowing up from the Gulf of Mexico fuels rain and thunderstorms all summer long. Details of these storms are in Chapter 10.

Figure 4-2:
Summer winds circulating around the Bermuda High.

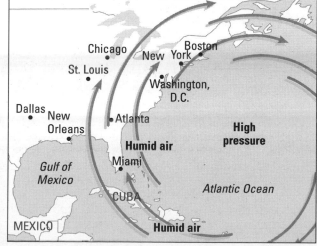

This same circulation of winds around the Bermuda High also helps drive the big clockwise circulation of ocean current known as the *Gulf Stream*. This pathway of wind and ocean current brings warm subtropical water up along the Eastern Seaboard. As Chapter 11 describes, this current of warm water invites tropical storms and hurricanes to stray up into the middle latitudes.

That cool Pacific High

The general circulation of the atmosphere helps account for the fact that even though summer temperatures often are hot in the western United States, the humidity usually is much lower and the summers are drier than in the East.

At about the same latitude as the Bermuda High in the north Atlantic, the north Pacific Ocean develops another large summertime pattern of high pressure. This is known as the *Pacific High,* or the Hawaiian High. Out West, the same circulation pattern has the opposite effect on humidity and summer rainfall as the Bermuda High has in the eastern U.S.

Winds circulating clockwise around the eastern side of the Pacific High carry cool and relatively dry air down from the more northern latitudes. This circulation helps drive a big current in the Pacific Ocean around in the same pattern. The combination of cold southward flowing water, known as the California Current, and the air moving above it bathes the West in cooler and drier summertime air. Check out Figure 4-3 for a picture of this pattern.

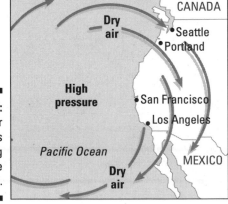

Figure 4-3:
Summer winds circulating around the Pacific High.

The winter lows

As the warmth of sunshine migrates south in the autumn, the high pressure systems in the northern oceans shrink and also head south. As they do, two low pressure systems slide down out of the far north and expand. Off the west coast of Canada, the big *Aleutian Low* will help generate many storms out of the north Pacific and sling them over the United States. Off the east coast of Canada, a low pressure area known as the *Icelandic Low* seems to be the attraction for many northeasterly storm tracks crossing the eastern U.S.

Bending the Winds

Three forces or motions make the winds blow horizontally, or sideways, the way they do. One gives the winds their power. Another slows them down near the ground. And another controls their direction in the upper atmosphere — *aloft*.

✔ First and foremost is *air pressure,* or more accurately, the *difference* between pressure one place and pressure another. Air moves outward from high and inward toward low pressure, and the bigger the difference between the two, the stronger the wind blows.

✔ The second force that shapes the wind is *friction*. It may not seem like it, but flowing air feels the effects of friction as it flows across a surface. You and I can't see it, but air acts somewhat like water flowing over a rocky river bottom. It gets slowed down and pushed around by everything it runs into.

✔ Third is the effect of Earth's *rotation,* which causes winds to bend, or change direction, from the paths they would follow under the influence only of pressure or friction. This effect steers winds to the right in the Northern Hemisphere and the left in the Southern Hemisphere and causes most storms in the mid-latitudes to travel from west to east.

Taking the pressure — it's a gas

Pressure is simply the air's weight. But because it is a gas, air's weight has dimensions that you don't encounter when you place your delicate little body on the scales. One big difference is that pressure is the effect of the weight of a fluid — a gas or a liquid — being exerted in all directions. It is air's weight from the top of the sky down as well as from the bottom up.

Measuring pressure: Barometers

When somebody uses the term *barometric pressure,* they are referring to the air pressure readings on the face of one of two types of barometers.

The *mercury barometer,* invented more than three centuries ago, is not so convenient, although it still is the more accurate of the two. This long glass tube is sealed at one end and filled with mercury, forming a vacuum so that no air can enter the tube. The other end of the tube is submerged in a small open pool of mercury, and inside the tube the heavy liquid settles down the tube at a level that depends on the outside air pressure. On average, at sea level, pressure causes the mercury to settle 29.92 inches up the tube.

A more portable and common barometer that does not use liquid is called an *aneroid* barometer. This instrument measures air pressure on a dial like a clock face marked off in inches and other units that correspond to the level of mercury in the tube. The pointer on the dial is attached to a flexible vacuum chamber that moves up and down with pressure changes.

Some barometers have *Fair* or *Stormy* or *Changeable* designations at different air pressure readings around their dial, but these benchmarks are not as helpful as you might think. The most useful information is not the momentary air pressure reading, but its *direction of change.* Falling pressure often indicates stormy weather is here or on the way, and rising pressure often means fair weather or clearing skies.

Try this at home: Hold your arm straight out from your side and feel the weight of it out there. All you feel is the weight of your arm. Luckily, you don't feel the weight of all the air overhead that is bearing down on the top of your arm. If you did, if you felt air's weight of 14.7 pounds per square inch only on top, you wouldn't be able to hold up your arm. This is the weight of one-inch by one-inch column of water 33 feet high. Instead of feeling this weight on your arm, at the same time the same weight (or pressure) is bearing down on you, it also is pushing up from underneath your arm.

At a particular place on the surface, pressure is a measure of how many molecules are in a column of air directly overhead. The air pressure changes from the bottom to the top. The force is greatest at the bottom because all the molecules are overhead. The higher up the column, the less the air pressure, because fewer air molecules are always overhead. At about 18,000 feet, about three and a half miles up, about half of the atmosphere's molecules are overhead and half are below you. So the air pressure is about half what it is at the surface.

Feeling the downward force of all of that atmosphere overhead, air molecules congregate more closely to one another near the ground than up in the sky.

In other words, the air is more dense near the ground. Vertically, or up and down, the differences in air pressure are huge compared to the differences horizontally, or sideways. In fact, the bulk of the air is so concentrated near the ground that a trip up an elevator in a skyscraper carries a passenger through greater differences in air pressure than would be measured across 1,000 miles of the surface. Between those big Hs and Ls on a weather map, the actual difference in air pressure may be only about 5 percent, and yet it is enough to move air masses and make winds blow hundreds of miles. Figure 4-4 illustrates how quickly air pressure changes with height.

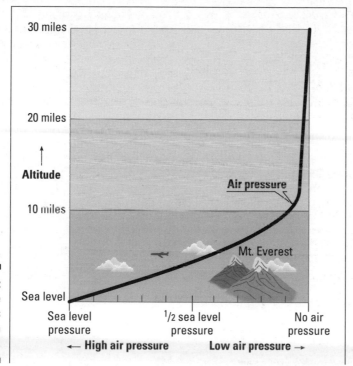

Figure 4-4: Air pressure changes quickly with height.

Here's the rub — total friction

If you and I could see air blowing over the land, it would look a lot like the behavior of water in a river. The motions of gases and liquids are a lot alike because they are both fluids. They both flow. In fact, my people at the Go Figure Academy of Sciences tell me that a lot of the important ideas about the behavior of flowing air were first tried out in dishpans of water.

Measuring winds

Weather forecasters don't get very far without knowing the direction and speed of winds. Some of science's oldest instruments — and some of its newest — are measuring these features of the atmosphere.

Wind vanes of all shapes and sizes have been pointing out its direction since ancient times. *Anemometers* that measure wind speed as it spins cups or propeller blades have been used for many years.

Twice a day, around the world, hundreds of weather balloons carrying instrument packages are set free to help gather some of the most important data for weather forecasting. They measure temperature, moisture, and the directions and speeds of winds at different levels in the atmosphere.

New Doppler radar devices called *wind profilers* have been designed to give weather scientists a vertical picture of the winds from the ground to an altitude of about ten miles. Aboard new satellites, laser beams are being used to measure winds.

When a parcel of air bumps into something while it is flowing, it slows down and finds a way around the obstacle before it continues on its way. And the air right behind that parcel gets slowed up and diverted from its preferred path. And likewise, the air parcel behind that gets diverted, and so on up the line as each parcel of blowing air encounters the turbulent effects of the obstacle. Figure 4-5 illustrates these effects on different layers of air.

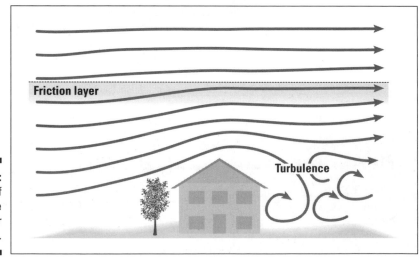

Figure 4-5:
The flow of air in the friction layer and aloft.

A commodious correction

It's time to get to the bottom of the question of the toilet bowl flush.

Winds blow around high and low pressure systems in the Northern Hemisphere in opposite directions from the way they blow around highs and lows in the Southern Hemisphere. And ocean currents also flow in opposite directions under the influence of the same forces. So, is it true, as urban legend has it, that toilets flush in opposite directions in the different hemispheres?

My people at the Go Figure Academy of Sciences have looked deeply into this profound question and have done the math. And the answer is No. Whatever is going on in the bowl, the times and distances involved are far too short to feel the effects of Earth's rotation.

All this commotion is caused by *friction,* and it has the effect of slowing down air in the same way that hitting your brakes slows down your car. The brakes depend on friction, and so does the tread on your tires. It's just that you can't see air's skidmarks. The effects of friction on air extend about two-thirds of a mile up into the atmosphere. Weather scientists refer to this as the *planetary boundary layer* or the *friction layer,* and the behavior of winds at this level are especially important to forecasters predicting weather for specific localities.

Air above this friction layer is called the *free atmosphere, the air aloft,* or the *upper atmosphere,* where the flows are not slowed or diverted nearly as much by the effects of friction between the surface winds and the landscape.

When forecasters refer to "the air aloft," most often they are describing conditions a mile or so up, where air pressure is lower and winds behave differently than they do at the surface. Without so much surface friction, they blow faster, and because they blow faster and over long distances, they react more strongly to the third big wind force, the effect of Earth's rotation, which changes their direction.

A perfectly straight curveball

Here is a Big Picture Question (BPQ): Why do the jet streams and all those big winds of the upper atmosphere over the middle latitudes mostly blow out of the west, and why does most of the weather travel from west to east across the United States? One way to get a handle on this BPQ is to play a little baseball.

Eye-balling the wind

Want to measure winds the old fashioned way? Here's a modified version of a scale first devised to help British sailors estimate wind speed.

Speed MPH	Name	Common effects
0–1	Calm	Smoke rises straight up
1–3	Light air	Smoke drifts
4–7	Light breeze	Feel it on your face and see leaves rustle
8–12	Gentle breeze	Leaves on the move, and flags wave
13–18	Moderate wind	Dust, leaves, and paper flies. Branches move
19–24	Fresh wind	Small trees sway
25–31	Strong wind	Large branches move. Whistles through wires
32–38	Gale	Trees sway. Hard to walk
39–46	Fresh gale	Twigs snap off trees
47–54	Strong gale	Branches break. Shingles blow loose
55–63	Storm	Trees break. Buildings damaged
64–72	Violent storm	Widespread damage
73–higher	Hurricane	Extreme damage

Suppose that I were a really good baseball pitcher, and you were a really good catcher. If I could perch myself up on the North Pole on a pitcher's mound that is high enough, I would see you out on the edge across the middle latitudes of the Northern Hemisphere. You give me the sign for a high fastball. Now here I go: I wind up in this great motion, and I throw you my fastest and straightest fastball. But look what happened: You missed it! I would swear that I threw you a perfectly straight pitch, and you would swear that it was a wild pitch that curved to the left clear out of the batter's box. The weird thing is, I am right, and you are right. Go figure.

This baseball game is taking place on a level playing field, as the saying goes, but the darn thing is spinning like a merry-go-round. From the North Pole, it's spinning counterclockwise. As far as the flying baseball is concerned, it followed a perfectly straight path since leaving my hand, but while it was traveling, you were moving off to its left. From behind the plate, the way you saw it, I take another look at the pitch on instant replay and sure enough, I have to agree with you that my fastball definitely tailed off to the left.

Figure 4-6 illustrates this effect from the vantage point of the ball and from the point of view of two people on a rotating sphere.

This curving to the right in the Northern Hemisphere and curving to the left in the Southern Hemisphere is the effect of Earth's rotation. In theory, at least, this deflection affects every ball that is thrown in the Northern Hemisphere, but I would not try to sell this idea to a coach if I were you. The truth is, if your catcher is missing your pitches from just 60 feet away, you're going to have to work on your fastball.

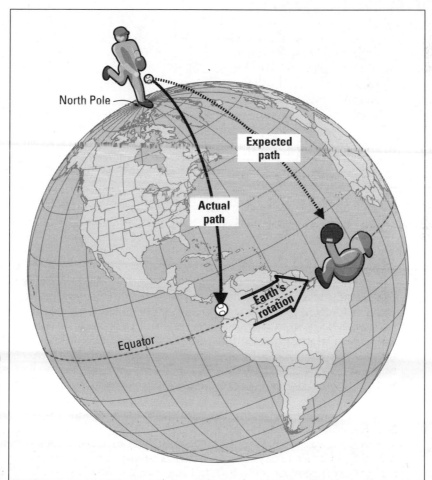

Figure 4-6:
The effect of Earth's rotation on a fast-traveling object.

North Pole

Expected path

Actual path

Earth's rotation

Equator

Pilots of transcontinental airliners, on the other hand, have to figure this effect into their flight plans. And Earth's rotation has this "bending" effect on the winds. This effect causes winds to circulate around high and low pressure systems the way they do. The effect of Earth's rotation is why winds in the Northern Hemisphere circulate clockwise around high pressure and counterclockwise around low pressure, and why in the Southern Hemisphere they flow counterclockwise around high pressure, and clockwise around low pressure.

The Winds Aloft

The upper atmosphere has a different look to it than the patterns you and I live with down near the ground. Everything aloft is sort of smoother. What looks like a strong high pressure area on a map of the surface becomes abbreviated into a *ridge* of high pressure in the atmosphere a few miles up, and likewise, a strong low pressure system on the ground is a mere *trough* in the air aloft.

The winds in the upper atmosphere still act the same way as surface winds in the sense that they want to flow clockwise over the high ridge and counterclockwise over the low trough, but in the air aloft the force of the Earth's rotation always is competing with these pressure forces and tending to straighten out their flows. As Figure 4-7 illustrates, the winds aloft are blowing much faster than they are near the ground, and their pattern has a large wavy look to it.

The Westerlies

In the simplest picture of the atmosphere, air is rising under the warming influence of the Sun near the Equator and falling toward the surface near the cold poles. If nothing else were happening in the atmosphere, the upper air would travel toward the poles, and surface winds would blow from the poles toward the Equator. But the atmosphere is not so simple, and the winds don't blow that way.

In the Northern Hemisphere, this flow of air toward the North Pole is being bent to the right by the force of Earth's rotation. The flow reaches a balance between the force pushing it north and the force bending it to the right. The result is a prevailing wind that blows high above the middle latitudes from west to east. This big band of winds is known as the *westerlies*.

The westerlies seldom flow in straight lines from east to west, but rather they follow a wavy pattern of large and persistent ridges and troughs. As the warming rays of the Sun fade farther south during winter, the westerlies migrate farther south over the United States and other regions of the Northern Hemisphere in response to changes in these big troughs and ridges. (Chapter 2 describes how the Earth's tilt causes the seasons.)

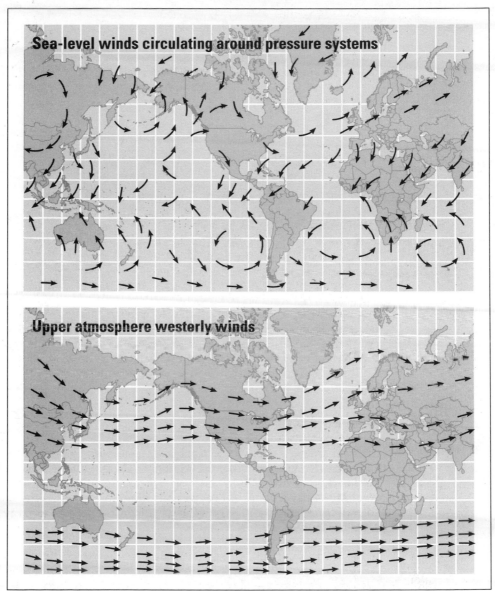

Sea-level winds circulating around pressure systems

Upper atmosphere westerly winds

Figure 4-7:
Typical
patterns of
wind and
pressure at
the surface
and aloft.

In response to the warming of spring and summer, the westerlies generally migrate farther north. Figure 4-8 shows you typical patterns of the westerlies during winter and summer.

A smoother pattern of westerlies develops in the Southern Hemisphere, because the air flows more uniformly over water, but the direction of the flow is the same. As the air flows from the Equator toward the South Pole, it gets bent to the left by the force of Earth's rotation.

The Jet Streams

Inside the upper atmosphere's westerlies, maybe eight miles to ten miles up, is an even stronger wind that blows right at the boundary between the cold air from the poles and the warmer air over the middle latitudes. This abrupt boundary of temperature — and sharply different air pressures — is called the *polar front* and extends from the upper atmosphere all the way to the surface. The wind up there is called the *polar jet stream*.

Polar jet stream

The *polar jet stream* is an upper atmosphere westerly wind — only more so. It is a name that is given to the fastest part of the westerlies. It typically flows about 100 miles per hour, although in winter the jet stream can kick up to speeds of more than 210 miles per hour as temperature differences in the atmosphere become greater.

Jet streams and the westerlies can make storms more powerful. They cause air to rush together and rush apart as they race through the upper atmosphere. This motion makes air move up and down between the upper atmosphere and the lower atmosphere and helps to ventilate the big storm systems.

Sometimes the jet stream is referred to as the storm track, but this notion does not very accurately explain what's going on up there. The jet stream is not really responsible for delivering a storm to a particular location. The upper winds are strongest where the air pressure differences are greatest — at the seam between two different air masses. And that seam is where storms develop. And certainly a powerful jet stream can add to the intensity of a storm. Think of it this way: The jet stream is overhead for the same reason the storm is in your face — two very different air masses are coming together.

Subtropical jet stream

In addition to the polar jet stream, other westerly jet streams develop in the upper atmosphere. One jet stream that develops farther to the south over the lower latitudes around 30 degrees north is the *subtropical jet stream*.

The subtropical jet stream occasionally delivers large amounts of moisture that fuels the winter storms over the southeastern United States. Crossing the Pacific Ocean, sometimes it turns northeastward and brings heavy rains to California and the Southwest. Chapter 9 describes these treacherous Pineapple Express storms.

Low-level jets

Another kind of jet sometimes flows near the surface in various regions and can have dramatic effects on local weather. Just a few hundred yards off the water, sometimes such a low-level jet races across the Pacific Ocean and blasts into the Coast Range mountains. Recent research shows that this jet causes some winter storms to practically burst with precipitation over the California coast as the wind drives into the Coast Range. Wind that flows up the Mississippi River Valley, fueling night-time thunderstorms, is sometimes referred to as a low-level jet.

The Tradewinds

Sailors of the world have long been acquainted with a feature of the general circulation of the atmosphere that this chapter describes earlier in the section "A World of Winds and Pressure." This global pattern creates a flow of air from the Equator that descends as high pressure at about 30 degrees north and south latitude. This descending air forms persistent high pressure areas in the oceans at these latitudes. The effect of Earth's rotation leads to clockwise wind circulations around high pressure north of the Equator, and counterclockwise wind circulations around high pressure south of the Equator. The flow at the surface around these high pressure systems is the tradewinds — to the right in the Northern Hemisphere and to the left south of the Equator, and so in both hemispheres from east to west.

This steady flow offers mariners a fairly reliable wind that will carry their vessels from east to west across the world's oceans. Christopher Columbus was carried by these winds across the Atlantic Ocean to the New World in 1492. His tradewind route became the regular route from Europe for trading vessels, and the winds became known as the *tradewinds*. Merchants would travel south from Europe to try to catch the easterly tradewinds across to the American colonies. On their return, they would follow the Gulf Stream to the north and try to catch the westerlies to carry them back.

Crossing the sea in sailing ships used to be a dangerous business. When they weren't concerned about violent storms, ship captains were worried about equally treacherous areas of calm air. The global circulation pattern that is illustrated in Figure 4-5 earlier in this chapter creates big areas of deadly calm at both ends of the tradewinds flow that earned dreadful reputations among sailing mariners.

Along the line of steamy storminess near the Equator, where the air pressure is pretty much the same from one place to the next, the band of quiet seas and calm winds became known as the *doldrums*. When you are "down in the doldrums," you are going nowhere.

Farther north in the Atlantic Ocean, the band of evenly distributed high pressure in the sinking air at roughly 30 degrees north — the Bermuda High — was an especially big threat to ships of yore. Caught in the calm for weeks, sailors and passengers could face starvation. Horses, which were common cargo on such journeys, often were thrown overboard, or eaten. The area came to be known as the *horse latitudes*.

A Scattering of Winds

Important regional and local winds are caused by air pressure differences created by conditions that change from season to season, from day to day, or even from day to night.

Coastal breezes

At the margins of the continents and along the shores of great lakes, daily breezes are regularly created by circulations formed by the different ways that water and land handle heat. Chapter 5 describes how warm air rises as it heats up. The land gets hot quickly and starts radiating the heat away. Water absorbs the Sun's heat more slowly, and keeps it much longer. Figure 4-9 illustrates how these different heat-absorbing and radiating properties affect the winds near the shore.

The day at the coastline typically begins with a land breeze, a wind blowing out to sea. The land has cooled more than the sea during the night, and its air pressure has become higher. The ocean is relatively warmer than the land, and the air pressure over it is relatively lower.

During the day, the Sun's rays quickly warm up the land and the air above it, causing the air to rise, forming an area of relatively low pressure. Over the water, meanwhile, temperature and air pressure stays pretty much the same. By afternoon, commonly, the air flow reverses. The land breeze becomes a sea breeze.

Valley and mountain breezes

Night and day often bring a reversal of breezes between valleys and mountains. As Figure 4-9 shows, the low-lying valleys warm up more intensely during the day, and the rising air flows up the slopes of the mountains. As night falls, the mountains cool more intensely, and this denser air sinks back down the slope as a mountain breeze.

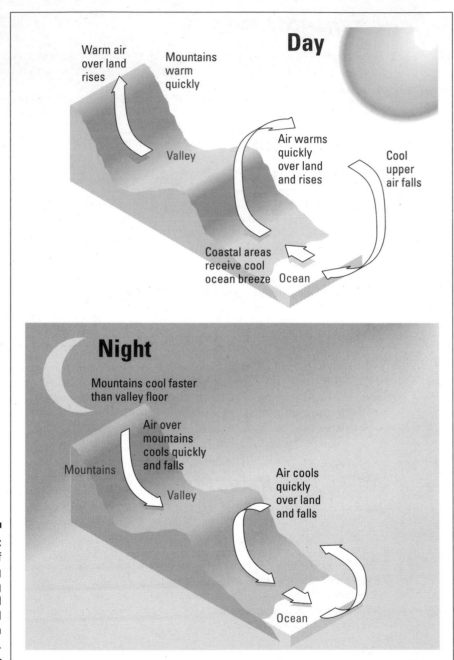

Figure 4-9:
The effect of
warming
and cooling
on coastal
and
mountain
breezes.

Asian monsoon

The regional wind of greatest importance is a seasonal pattern that delivers rainfall that is the lifeblood of food production for the multitudes of South Asia. Failures of the Asian monsoon have been human disasters of staggering proportions in history. A *monsoon* is a wind that changes directions with the seasons.

The winter monsoon brings fair skies and dry weather to India and South Asia. High pressure develops over the cold regions of Siberia, creating big dry northeasterly winds that blow from the land out to the sea.

The summer monsoon brings heavy rains. The Asian continent warms up, creating low pressure that draws in toward the land big flows of moist air from the Indian Ocean and the South China Sea.

Southwestern monsoon

A similar but less dominant monsoon pattern develops in other regions of the world, including the southwestern desert of the United States. In midsummer, the warming desert creates rising air and low pressure. This brings moist subtropical air up from the Pacific Ocean and the Gulf of California that often fuels thunderstorms and showers, and sometimes flash floods.

Chinooks

As westerly winds of winter and spring encounter a north-south mountain range such as the Rockies, they push air up the westerly slopes, often causing condensation and rain or snow. As it reaches the mountain top, this air is drained of its moisture, begins sliding down the eastern side of the mountain, warming and becoming more dense as it falls. From Canada to the southwestern desert, these winds are called *Chinooks*.

These strong, sometimes violent winds race down through the foothills of the eastern Rockies. By the time they reach the cold, snow-clad Great Plains, Chinooks can cause shocking rises in temperatures. Within minutes, temperatures have been known to climb more than 40 degrees. Chinook winds sometimes are called *snow eaters*.

Santa Anas

One of the most dangerous winds in the world is named after the Santa Ana Canyon, although its causes and effects go far beyond that Southern California canyon. Far to the north and east, cooling temperatures over the Great Basin during fall and winter develop a large area of high pressure and sets in motion a big clockwise circulation of dry winds. Warming as they fall from the high plateau over the southwestern desert, the Santa Ana winds can charge violently through the passes and canyons of the San Gabriel and San Bernardino mountains.

The desert brushlands are tinder-dry, and when the Santa Anas are blowing, the slightest spark can grow to a raging firestorm in a matter of minutes. Such wildfires have claimed many lives and caused billions of dollars in property losses in the last several years. Such giant high pressure circulations in the inland west can generate treacherous Santa Ana-type winds over a wide area of California and the Southwest.

Haboobs

Whirlwinds or dust devils are common sights and usually small and momentary features in big deserts across the world, where hot, rising winds begin spinning in local turbulence. But especially violent thunderstorms over sandy deserts can generate big, damaging winds known as *haboobs*. These winds derive their name from the Arabic word *habb,* for a wind caused by a downdraft.

Tons of sand and dust can become airborne by the downdraft, or downward charging wind, along the leading edge of a big thunderstorm. The dense, dark cloud can extend across the landscape for many miles and can completely engulf a desert town or city. Haboobs are most common in the Africa Sudan and the southwestern desert of the United States.

Chapter 5

Getting Cirrus

*1*s rain on the way? Will there be a break in the storm? When you and I look to the sky for signs of change in the weather, the objects of this inspection are clouds. Without really thinking about it, experience tells you what to look for. High, white, wispy strands up against the blue sky aren't carrying anything that's going to land on your head. But the dark, low mass of gray has the look and feel of rain. You may not know them by the names they are called, and you may not know why they are there, but already you know a lot about clouds — more than you realize. As you compare the names and descriptions of the clouds in this chapter with the color photographs of those clouds at the center of this book, after awhile you may begin to feel like someone attending an old class reunion. No, you say to yourself, the name doesn't ring a bell, but that face is sure familiar!

They are the meat and potatoes of weather, these remarkable formations we call clouds. You and I know this fact without even thinking about it: Not a drop of rain or a snowflake or a single hailstone will fall from the sky unless first a cloud is formed. *But how is a cloud formed?* This chapter answers that question. And the names and descriptions of the ten main types of clouds are spelled out, along with their importance as forecasters of changes in weather.

Clouds are the beauties of the sky. Learning their names is a little like becoming acquainted with a few great master painters. Every day their canvas is up there for you to view, and most every day you can watch them work. Telling one kind of cloud from another is not really essential, of course. But it's fun, and sometimes it's especially rewarding. The sight of the setting Sun is quite nice, naturally. But the one I remember most is not of the Sun itself, but what happened to its fading rays on the undersides of the high clouds that I hadn't noticed before. The Sun already was over the horizon when the *cirrus* clouds took its last light and brilliantly illuminated the sky with rose and orange and gold.

Making Clouds: The Heavy Lifting

Clouds form when condensation takes place, a process that Chapter 3 describes in connection with precipitation. Clouds are where all three phases or forms of water are commonly in action: gaseous vapor, solid crystals, and liquid droplets. When it comes to the kinds of weather that you and I worry about most, clouds are where the action is. The more you know, the better sense you have of what to expect. You may begin to recognize the progression of changing cloud patterns in the sky that can signal hours ahead of time the advance of a distant warm front storm.

If you are going to make a cloud, you are going to have to cool air to its *dewpoint* — to the temperature where it is saturated with water vapor, and the vapor begins condensing into ice crystals or water droplets. That's what a cloud is — ice crystals or tiny water droplets, or a combination of both. The best way to cool air is to get it off the ground and move it up into the realm of lower air pressure, where it expands and its temperature drops. One way or another, you're going to have to lift the air.

Heating up

Just as warm water hangs around the surface of a swimming pool or an ocean, warm air floats upward through cooler air. The warmer air, like the warmer water, is less dense — more buoyant.

This *buoyancy* tendency of a lighter fluid to float upward through a heavier fluid is the basis for the powerful and dramatic cloud-making process known as *convection*.

The vertical *cumulus* clouds — often called *convective* clouds — that bring showers to summer afternoons most often are spawned by convection, by air rising from the Sun-warmed surface, reaching its dewpoint and condensing. The cooler the air, the less water vapor it can hold. Soaring birds such as hawks and vultures are experts at riding the warm updrafts known as *thermals* created by local convection.

Convection also is an internal process that takes place inside a forming cloud in addition to whatever mechanism first gave the air its lift. Water condensing from its gaseous vapor form into ice crystals or water droplets gives off heat. This warming from latent heat, which Chapter 2 describes, gives the air another boost upward.

Crowd control

When air moves from one place to another across the landscape, other air moves out of its way. When air converges from more than one direction onto a common location on the surface, the atmosphere has a crowd control problem on its hands. This problem happens all the time — for example, when air circulates counterclockwise and *inward* toward a low pressure system in the Northern Hemisphere. The atmosphere needs to find a place for the extra incoming air to occupy. And across the surface, of course, the only direction to go is *up*.

Convergence can lift an entire layer of air hundreds of miles across, although its motion is more gentle than other mechanisms that give air a lift. In response to the convergence of air near the surface, the whole atmosphere can bulge upward. The high, thin veil of cloud cover known as *cirrostratus* can be formed by convergence.

Frontal assaults

When warm and cold air masses do battle, as Chapter 2 describes, the atmosphere is going to get a rise out of it. Air is rising all along the advancing fronts as big winter storms march from west to east across the United States and other regions in the middle latitudes. But there are important differences in the clouds between a battle that is won by a cold front and one that is won by a warm front.

A *cold front* is a relatively fast-moving creature, typically traveling across the landscape at around 30 miles per hour. As Figure 5-1 illustrates, the blunt nose of this dense air vigorously lifts retreating warmer air at the boundary of the two masses. Vertical clouds form at the boundary, quickly developing rain or winter snowfall and occasionally spawning thunderstorms.

An advancing *warm front,* by contrast, is a more gently sloping slowpoke, typically moving along at maybe ten or 20 miles per hour. As Figure 5-2 shows, this warm air gradually rides up over the top of the retreating cold air. Thickening layers of clouds form far ahead of the frontal boundary and drop precipitation that is usually slower, steadier, and more widespread.

Over the top

Encountering a barrier such as a mountain range that it can't go around, the air rides up its slopes, cooling toward its dewpoint as it rises. Clouds formed by this process are called *orographic,* which is worth mentioning only because once in a while you might hear a meteorologist refer to *orographic clouds* or an *orographic storm.*

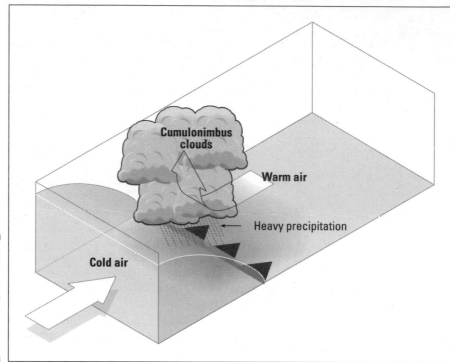

Figure 5-1:
Cloud
formations
caused by
the advance
of a cold
front.

Orographic lifting is an important process to the weather of the western
United States. The north-south mountain ranges of the Cascades, the Sierra
Nevada, and the Rockies form great, up-lifting barriers to the flow of air arriving from the west. The citizens of California depend for most of their water
supply on the rain and snowfall in the Sierra Nevada, and much of it is generated by the mountains giving an extra *orographic lift* to incoming Pacific
storms. The height of clouds generated by this motion often range far above
the mountain tops.

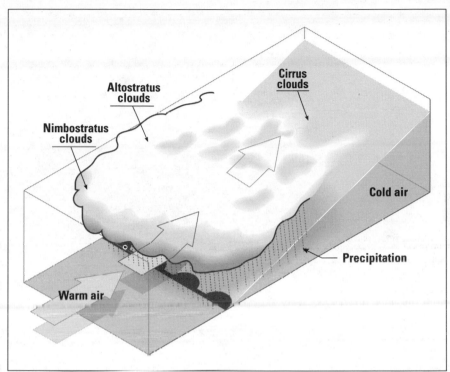

Figure 5-2:
Cloud
formations
caused by
the advance
of a warm
front.

As Figure 5-3 illustrates, the up-lifting process on the *windward* side of the
mountain (the side that faces the prevailing winds) produces another interesting weather phenomenon on the *leeward* side of the ranges, which is out
of the wind. Just as the up-lifting makes for especially rainy or snowy conditions on one side, it produces what is known as a *rainshadow,* or especially
dry conditions, on the other. For example, the desert known as Death Valley,
one of the driest places on Earth, lies within the rainshadow of the snow-
capped Sierra Nevada.

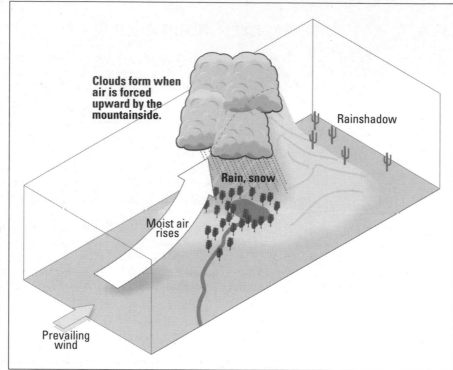

Figure 5-3:
Mountain ranges lift air and make storms on their wind-ward side.

A Question of Stability

Making a cloud without getting air to rise is hard, but something else is even more essential. The most important thing that has to happen for a cloud to form is for the temperature of the air to fall to its dewpoint or frost point — that condition called saturation, when it holds all of the water vapor that it can and not one molecule more. As it cools beyond this point, it causes *condensation,* and a cloud forms.

This process of rising takes place in a way that is not so obvious. A lot depends not only on the air that is on the rise, but also on the temperature and other features of the air it is rising into. You will hear a forecaster use the term *unstable air* or *unstable conditions* to describe the circumstances that lead to the greater development of clouds and increased likelihood of precipitation. What the forecaster is getting at is this:

Inversions: Putting a lid on it

What weather forecasters think of as the stability of air has to do with the temperature of the atmosphere at different heights — its temperature profile. They know that a rising parcel of warm air will cool at a certain rate. They compare that rate of cooling with the temperature profile of the atmosphere and come up with an estimate of its stability — the likelihood that it will create clouds and precipitation.

Unstable air has a temperature profile that is always cooler than a parcel of air rising through it. Stable air has a temperature profile that is close to the standard cooling rates of rising air, and so it discourages this ever-upward flow. And then there is the other side of the coin — air with a temperature profile that is warmer rather than cooler above the surface.

These so-called temperature "inversions" really put a lid on rising air. Not only do they discourage cloud formation and storms, these conditions are very unpopular with city dwellers and farmers in some parts of the country.

When a region is socked in with fog day after day, you can be sure a temperature inversion is overhead. Above places like Los Angeles, when air can't rise and mix, it doesn't take long before it gets pretty thick and stinky. On a farm where winter crops grow, an inversion means that colder air is hanging down around the tender crops, threatening to damage them with frost. That's what all those smudge pots and big fans that look like airplane propellers are for — to ventilate the fields. Hoping to prevent freezing, farmers are trying to get that trapped cold air to mix with the warmer air above the ground.

Under some circumstance, the air is particularly prone to keep on rising and cooling and condensing into cloud long after the boost that sent it upward in the first place has petered out. It continues to float above the air around it — in other words, to remain *buoyant.* As air rises, it moves into lower pressure, and so it expands. As it expands, it cools. The air cools at a constant rate of 5.5 degrees Fahrenheit for every 1,000 feet it rises. When conditions are unstable, this rate of cooling still keeps it warmer than the air around it, and so it continues to rise until it reaches saturation, its dewpoint. As it condenses into cloud, it gets another boost upward. The process of condensation, which Chapter 3 also describes, converts water from gas vapor to ice crystals or tiny liquid droplets. Condensation releases *latent heat* into the atmosphere. This particular parcel of air is giving back to the atmosphere the heat that was taken when the liquid water evaporated into gaseous water vapor in the first place. Condensation gives the air new warmth and still more buoyancy.

The rate of cooling of the rising air remains the same, but the vertical *temperature profile* of the surrounding air varies from day to day. That is why forecasters go to such trouble to find out what is going on with temperatures and other conditions in the upper atmosphere — the air aloft. This is a big reason why releasing those hundreds of weather balloons twice a day is worth the time and trouble, and the expense. Knowing the rate that air cools as it rises, and knowing the temperatures of the upper air through which the rising air is moving, a forecaster gets a pretty good handle on what to expect by way of cloudiness and precipitation.

When the air aloft is much colder, even as the warmer air rises and cools, it remains warmer than the surrounding air, and so keeps on rising. The higher it can rise, the more of its water vapor can condense into clouds, and the taller the clouds can grow.

Clouds by Class

No two clouds are exactly alike, it is true, but they aren't all entirely different from one another either. You don't have to be a meteorologist to recognize that clouds come in certain styles. As a rule, one cloud in the sky looks somewhat like the one next to it, because they are made by the same process. Knowing these different types and the different processes that make them is very helpful to forecasting weather and to getting a handle on what is going on up there. Being able to tell one type of cloud from another is always fun, and once in a while it can be valuable personal safety information.

British naturalist Luke Howard devised a system of classifying clouds 200 years ago that still is used today. It is a simple system, although it sounds a little strange because of the funny old Latin words it uses.

The clouds are labeled according to their appearance and according to their altitude. But, really, when you look at them, there are only three main types of clouds — stringy, heapy, and layered. And there are only three altitudes — high, middle, and low. And then there's one more important cloud type — the big, heaped clouds that generate vertically into the sky. Some of these vertical clouds are so harmless that they are called *fair weather clouds*. But others are the monsters that bring the world its most dangerous and violent weather.

So, how's your Latin? A little cloudius on your Latinus? Relax — all you need to know is this:

- ✔ *Cirro,* in the names of high clouds, means "a curl of hair."
- ✔ *Alto* means "middle."
- ✔ *Stratus* means "layer."
- ✔ *Cumulus* means "heap."
- ✔ *Nimbus* means rain.

If you really want to know, the classes of high, middle, and low clouds are called *stratiform* clouds because they form in horizontal layers. The vertical clouds, called *cumuliform* clouds, form tall heaps. Figure 5-4 illustrates the main cloud types.

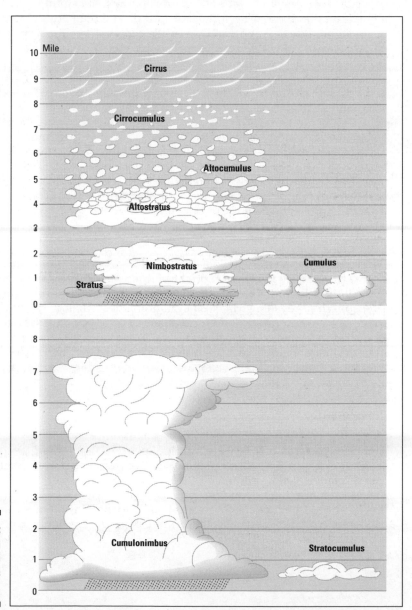

Figure 5-4: Main cloud types and their heights in the sky.

High clouds

The atmosphere is higher over the warm Equator and lower over the cold poles, so the altitude of cloud types varies with latitude. (Check out the color pages of *Weather For Dummies* for photos of the main cloud types.) Keep in mind, also, that these descriptions of altitude and temperature are not written in stone. One day to the next, one place to another, the atmosphere has its own way of doing things. So even though specific numbers are given for cloud heights, for example, take them as a guide.

In the middle latitudes, where most people live, the bases of high clouds form above about 23,000 feet, more than four miles up. The air at this altitude is cold, well below zero, as a rule, and these generally thin *cirro* clouds are composed almost entirely of ice crystals. You can see icy brightness in the patterns of even the puny cumulus forms at this altitude. All these clouds live in a region of high winds. These clouds are not weather-makers, for the most part. They cause no precipitation themselves, although cirrostratus may be a harbinger of rain or snow. They seldom cover the entire sky and are not thick enough to prevent sunlight from casting shadows.

Cirrus

Cirrus clouds are delicate-looking, thin, and wispy strands that are silvery and almost transparent against the bright blue sky (see Figure 5-5). Cirrus clouds in the middle latitudes are commonly carried by the westerlies, the upper atmosphere's prevailing winds that Chapter 4 describes. Often, they take the shape of a concentrated cloud form attached to a long, thin strand. These long streaks, sometimes called mares' tails, are ice crystals that have fallen from the "parent cloud" and are trailing downward into winds that are slower than the winds at the higher elevations where they formed. The appearance of *cirrus* can mean that bad weather is on the way.

Cirrocumulus

Cirrocumulus clouds are thin patches of small, white rounded clouds arranged usually in patterns of long waves or rippling rows that remind people of fish scales (see Figure 5-6). Cirroculmulus are the clouds that are sometimes called "mackerel sky," although these patterns rarely cover the entire sky. The clouds form under conditions of *wind shear,* when winds change direction or speed from one height to another. Cirrocumulus often are seen as signals of precipitation ahead, because wind shear occurs out in front of advancing storms.

Jim Reed.

Figure 5-5:
Cirrus
clouds.

Figure 5-6:
Cirrocumulus
clouds.

Jim Reed.

Cirrostratus

When you see a halo around the sun or the moon, you are looking through a thin, silvery veil that is a *cirrostratus cloud* (see Figure 5-7). As Chapter 12 describes, sunlight and moonlight scatters, or is *refracted,* as it passes through this nearly transparent lens of ice crystals. Thin sheetlike cirrostratus clouds often spread over the entire sky. Thickening cirrostratus clouds may signal the onset of rain or snowfall in the next 12 to 24 hours, as Chapter 2 describes, because they often form ahead of a storm generated by an advancing *warm front.* The veil of cirrostratus will become progressively thicker and then be replaced by lower, denser cloud types.

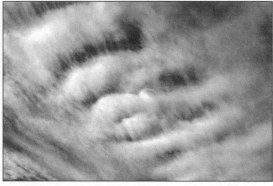

Figure 5-7:
Cirrostratus
clouds.

National Center for Atmospheric Research/University Corporation for Atmospheric Research/National Science Foundation.

Middle clouds

In the atmosphere between 6,500 feet and 23,000 feet — from a little over four miles up down to about a mile and a half — are clouds composed of both water droplets and ice crystals. Temperatures in these clouds range from about 32 degrees to -13 degrees, and the water droplets they contain are *supercooled.* These droplets are made up of pure water and, because of their tiny size, are able to remain liquid rather than freeze into ice even though the temperature is well below water's normal freezing point. The middle *altos* clouds are generally thicker than the high *cirros,* but they are seldom more than about a half-mile from bottom to top.

Altocumulus

Altocumulus clouds are gray, puffy patches that sometimes form wave patterns or bands (see Figure 5-8). Their patches are bigger than cirrocumulus, and they are likely to cover more of the sky. Also, their edges are more sharply defined because they contain water droplets rather than ice crystals. Rising air may cause them to form "little castles." They rarely cause precipitation that reaches the ground, although they indicate the presence of a layer of unstable air aloft. Altocumulus form by convection, the lifting of air, and may signal the advance of a cold front. If they appear in the morning of a hot, humid summer day, conditions may be right for thunderstorms in the afternoon.

National Center for Atmospheric Research/University Corporation for Atmospheric Research/National Science Foundation.

Figure 5-8:
Altocumulus
clouds.

Altostratus

This layer of gray or bluish-white ice crystals and water droplets covers the entire sky much more thickly than the white veil of cirrostratus. *Altostratus* may be thin enough to let the sun or even the moon peep through as a barely visible "watery" disk, but they are too thick to form halos or to allow shadows to form on the ground (see Figure 5-9). These cloud layers often arrive ahead of widespread rain or snow.

Figure 5-9:
Altostratus
clouds.

National Center for Atmospheric Research/University Corporation for Atmospheric Research/National Science Foundation.

Low clouds

The bases of these clouds extend from the ground, in the case of fog, up to about a mile and a half above the ground. Chapter 3 describes the details of fog. Low clouds usually are made up of water droplets, although during winter they may contain ice crystals or snow.

Stratocumulus

These large lumps of low clouds, called *stratocumulus,* appear in dark rows or patches (see Figure 5-10). They are thick gray clouds, but their irregular roll pattern may be broken by patches of blue sky and by dramatic, streaking rays of sunlight. They look threatening, but stratocumulus clouds rarely bring precipitation.

Figure 5-10: Strato-cumulus clouds.

Jim Reed.

Here's a simple rule that helps distinguish altocumulus from stratocumulus: Extend your arm up toward the clouds. Individual altocumulus are about the size of your thumb, while stratocumulus about the size of your fist.

Stratus

If you're in a fog , you are in a *stratus cloud* that is on the ground (see Figure 5-11). This uniformly gray cloud cover extends from horizon to horizon. It is the cloud of lifted fog, the gray overcast of the early morning seashore. Stratus may block out the tops of buildings and hills and occasionally bring mist or drizzle. But stratus is not a rain cloud.

Nimbostratus

Nimbostratus are rain clouds (see Figure 5-12). Nimbostratus is a gray deck of cloudiness that is darker than a layer of stratus and more ragged in pattern and less uniform at its base. The Sun is completely blocked out of the sky. This low, dark, full-sky cloud cover produces long periods of light to moderate rain or snowfall. Because of the precipitation and foggy conditions, the out-lines of these clouds can be pretty vague. The underside of a nimbostratus may be marked by drifts of lower ragged cloud fragments called *scuds.*

Figure 5-11:
Stratus
clouds.

National Center for Atmospheric Research/University Corporation for
Atmospheric Research/National Science Foundation.

Figure 5-12:
Nimbo-
stratus
clouds.

Jim Reed.

How high is high?

You probably know this already: Estimating the height of something in the sky is not so easy.

Airports and weather service offices have instruments that help them measure the distance from the ground up to the base of the clouds — what they call the *cloud ceiling.* Most often, they bounce a beam of light off the clouds, which is seen by a detector on the ground. They know the angle of the beam and the distance to the detector, and then they do the math.

For the rest of the population, it's a matter of making very rough estimates and using whatever features of the landscape may be handy.

You can use mountains and hills and tall buildings as yardsticks to the height of clouds. Sometimes airplanes can help. Small planes are usually flying below 12,000 feet, while big airliners are cruising up there at 33,000 feet or so.

As a person gets better and better at identifying the types of clouds, a funny thing happens to the problem of estimating heights. It sort of gets turned on its head. Instead of wondering how high those clouds are, you'll say to yourself: "Those babies are *altocumulus,* they must be about 15,000 feet!"

Vertical clouds

The clouds that develop vertically in the atmosphere, the *cumuliforms* that form from bottom to top, behave very differently than those *stratiforms* that occupy the sky in separate levels. The clouds of the high, middle, and lower levels are born out of air that is rising upward at the speed of roughly one mile an hour. Vertical clouds are charging bulls by comparison. Air can surge up through some of these giants at more than 100 miles an hour.

Vertical clouds are sometimes called *convective* clouds because they owe their existence and their sometimes explosive growth to the rapid vertical mixing of warmer and colder air. Often, they form by air radiating upward as the Sun's heat energy warms the land through the day. Stable conditions in the upper air keep a lid on cumulus clouds, but in unstable upper air, these clouds can grow to great heights in a matter of minutes.

Cumulus

A warm and otherwise clear afternoon may see the fleeting appearance of *cumulus clouds,* which are big, bright, sharply defined white pillows with darker undersides and flat bases (see Figure 5-13). Partly cloudy skies in fair weather often are populated by these "fair weather cumulus." These clouds form in updrafts of air and are encircled by downdrafts of air, marked by clear sky. Individual clouds commonly last less than an hour.

Figure 5-13:
Cumulus
clouds.

National Center for Atmospheric Research/University Corporation for Atmospheric Research/National Science Foundation.

Convection, or the rising of warm air, is the key to cumulus. They will outline the warm boundary around a cool lake, for example, or even the banks of a large river. They will originate and grow as the day's heat builds and dissipate as it cools. As the Sun's heat reaches its maximum by late afternoon, so, too, will the cumulus clouds reach their greatest heights.

A more potent type of cumulus cloud called *cumulus congestus* takes shape in warm, humid, often unstable conditions. These clouds can quickly build into closely packed, high vertical "towering cumulus" that bring scattered showers. If they continue to build, cumulus congestus can become giant cumulonimbus thunderstorm clouds.

Cumulonimbus

In a matter of 15 minutes, the cauliflower tops of a cumulus congestus cloud can surge high into the atmosphere, forming a flat, anvil-shaped crown reaching eight or ten miles altitude. As the top of the *cumulonimbus* pokes into the lower reaches of the stratosphere, it encounters high-speed winds that give its thunderhead that swept-away look (see Figure 5-14). The advance of this deck of cirrus often marks the approach of a cumulonimbus cloud, and then you might feel a gust of cool air from its downdraft just before a thunderstorm.

Figure 5-14:
Cumulonimbus clouds.

National Center for Atmospheric Research/University Corporation for Atmospheric Research/National Science Foundation.

These enormous *cumulonimbus clouds* can form individual storms or be part of a line of towers called a *squall line.* These clouds can bring lightning and torrential downpours of rain and hail. They can spawn violent downdrafts and tornadoes. Nobody who has experienced the weather violence that can be generated by the biggest of these monster clouds, called *supercells,* is likely to mistake them for anything else. Chapter 9 describes these thunderstorms in detail.

Special clouds

Always the atmosphere seems to be ready to play tricks on the likes of you and me with the clouds that take shape. Even if you are a veteran cloud-watcher, once in a while, an especially strange shape might stop you in your tracks and make you wonder, "What in the world is going on up there?"

Here is a rundown on some of the strange shapes you might see:

Lenticular

Mountains can cause moist air to form waves as it flows over their tops. For miles downwind of the mountains, these so-called *standing waves* hold their shapes, even as winds flow through them. Air flowing up the ascending crests of these waves can condense into clouds that form flat lens shapes called *lenticular clouds* (see Figure 5-15). These strange forms have been responsible for more than a few reports of UFO sightings.

A cloudy picture

You and I naturally think of clouds as creatures that change the look of one day from another. But the people in the climate wing of the Go Figure Academy of Sciences look at clouds from a different point of view. They're interested in the Big Picture — not in the clouds of storms that forecasters care about, but rather in the *total cloud cover* over the Earth. They want to know whether this total is going to be increasing or not, and what effects such changes will have on the future climate.

They think they know that global warming, which Chapter 11 describes, will bring about more cloudiness because it will increase evaporation and throw more water vapor into the atmosphere. But the effect of this cloudiness on climate is one of the cutting-edge debates among the Big Picture people.

Among the questions: What kinds of clouds would they be? More *cumulus* clouds are likely to absorb more heat radiating off the surface, making the warming warmer, and they think the same effect would come from more high-level *stratiform,* or layered, *cirrus* clouds. On the other hand, more middle-level *altos* or low-level *stratus* clouds could have the opposite, cooling effect.

Figure 5-15:
Lenticular
clouds.

National Center for Atmospheric Research/University Corporation for Atmospheric Research/National Science Foundation.

Billows

Layers of different air temperature can develop waves along their common boundary, and the *billow clouds* that form in these waves can look very much like a row of ocean waves.

Banner

A cloud that forms near the top of an isolated mountain peak and drapes away downwind is called a *banner cloud.*

Pileus

Moist air can get pushed over the top of a developing cumulus tower, forming a fuzzy cap cloud or *pileus,* which is, of course, Latin for "skullcap."

Mammatus

Unlike most clouds that form in rising air, *mammatus clouds* take shape as big bags of sinking air under cumulus or cumulonimbus clouds (see Figure 5-16). These big, dark sinking bags often appear ominous. Look for signs of rotation in these formations. If you see rotation, it may be a sign of a tornado. If there is no rotation, it is a harmless mammatus. In fact, mammatus often signals that the worst is over. The storm is drying up and cooling off. The air is cooling as it is sinking because it is giving up more heat through evaporation of water droplets than it is gaining through its declining altitude.

Figure 5-16:
Mammatus
clouds.

National Center for Atmospheric Research/University Corporation for Atmospheric Research/National Science Foundation.

Contrails

The exhaust from a jet aircraft often forms a line of cloud as water vapor is blown into the cold surrounding air. The shapes of these streaks of condensation can be clues to the upper atmosphere's winds and moisture content. Strong winds will quickly spread the streak. If *contrails* do not rapidly evaporate, it means that the air up there is relatively humid.

Clouds of the stratosphere

Almost all of the water vapor and almost all of the clouds occupy the lower eight miles of the atmosphere known as the *troposphere.* Occasionally a giant cumulonimbus cloud or a lenticular cloud over an especially high mountain will reach higher into the *stratosphere,* but weather scientists don't very often look for clouds up there. Still, once in a while, a rare cloud is observed in the upper reaches of the sky.

Nacreous

Over polar regions, at an altitude of about 20 miles, silky *nacreous,* or mother of pearl, clouds occasionally are illuminated by the long nights of the winter Sun. Weather scientists think they are made up of ice crystals or supercooled liquid water.

Noctilucent

Far above the layer of the atmosphere responsible for weather, some 50 miles up, a rare and mysterious layer of wavy bluish-white clouds is revealed long after sunset. Because of their very high altitude these clouds, composed of tiny ice particles, strangely illuminate the night at high latitudes (see Figure 5-17). *Noctilucent clouds* means "luminous night clouds."

Figure 5-17: Noctilucent clouds.

National Center for Atmospheric Research/University Corporation for Atmospheric Research/National Science Foundation.

Chapter 6

Climate Is What You Expect; Weather Is What You Get

At the Go Figure Academy of Sciences, the people who work in the climate wing of the meteorology building have their own ways of looking at weather. They are not as interested as forecasters in day-to-day weather events in particular places. It's the longer term view of weather on a somewhat bigger scale that gets the attention of the climate people. You know, it's the vision thing, The Big Picture. Weather is a tree, you might say, and climate is the forest.

In this chapter, I tell you everything you need to know about climate — why it's important, what (or who!) El Niño is, and why you should care. Before you know it, you'll be seeing the Big Picture, too. You won't even need a meteorology degree, and you'll be thinking like a climatologist — if you really want to!

Climate or Weather?

The key difference between climate and weather is time frame. People who are talking about individual storms are talking about *weather*. *Climate* is the pattern of weather over the duration of a season or a year, a decade, or even longer. Weather conditions tell you whether or not a storm is coming. Climate conditions might tell you something about sea surface temperatures or the shape of the jet stream, the storm track, and so where storms are likely to go during a season.

Climate or weather — so what? Is this difference important to you? Sometimes it is, sometimes it isn't. Here are some reasons why it might be:

- You want to know what you are talking about.
- You want to know how to make use of the information. (For example, if someone tells you a hurricane is on the way, it might be time to head for the hills. If someone tells you El Niño is on the way, it might be time to put on a new roof.)
- You can be sure that a farmer knows the difference. Weather information tells him how his crop is going to grow. Climate information tells him whether he should plant that crop in the first place.

Climates of the World

Climate is pretty basic information about a place. Technically, *climatology* is the average of a region's seasonal weather data, but in those dry statistics — and the wet statistics — is information that really tells you a lot about a place. It tells you if you want to live there, for one thing, and if you do, what kinds of clothes you'll wear and what type of house you will live in. There are several ways experts describe different climate features of a place, but the basics are, well, basically the same. Most often, a region's climate is classified according to these two features: first, the average amount of precipitation — rain or snow — that falls during the year and its variation from season to season, and, second, the range of average high and low temperatures during a typical day and their seasonal variation.

The "climate" climate

Try this at home. Notice how often people use the word climate — on television or in the newspaper or wherever — and how seldom they actually mean climate. You hear it fairly often in news accounts or political commentaries about this or that. But almost always the word is used figuratively. That is to say, they are not really talking about average seasonal characteristics of a place or about large ocean and atmospheric features that affect patterns of weather over time.

For example, there is the popular Economic Climate and the Business Climate. These climates are sometimes favorable, but usually they are unfavorable, and always they could be improved. There is another popular climate on newspaper opinion pages, and that is the Political Climate. When this comes up, usually the Political Climate is going from bad to worse. Another favorite is the Climate of Fear — which is never good.

My people at the Go Figure Academy of Sciences have been trying to look into this, but it's not easy. Maybe you have noticed this, too. It's hard to find one of these opinionated commentators when you really want one. Maybe the climate just isn't right.

Every place on Earth has its own climate, its own kinds of winters, springs, summers, and falls, more or less, and it has been climatologist's job to describe what "normal" precipitation or temperature patterns are at different times of year here and there. A lot of cataloging of weather statistics has been going on. Experts refer to these sets of statistical averages as a *climatology* of a place. Climatology tells you what to expect of one season or another in a particular locale.

Here are the basic climate types:

- ✔ **Tropical climates:** In the regions around the Equator, all year long the Sun is high in the sky. There is really no winter, no cold season between the Tropics of Capricorn and Cancer. Often it is humid, rainy, and always warm. These conditions nourish the world's tropical rainforests and jungles. Tropical climates stretch across Africa south of the Sahara Desert and across South America over the vast Amazon region of Brazil.

- ✔ **Dry climates:** The deserts of the world and the semidry grasslands known as the *steppes* occupy 26 percent of the land area. In dry climates, most of the year chances are good that more water will be lost to evaporation, from the ground to the sky, than will be gained as precipitation, from the sky to the ground. North America's dry climates include the deserts of northern Mexico and the southwestern United States, the strip along the eastern slopes of the Sierra Nevada, and the Great Basin. The semidry climates in North America occur in coastal southern California, the northern valleys of the Great Basin, and in most of the Great Plains.

- ✔ **Mid-latitude mild-winter climates:** These subtropical climates are often influenced by ocean conditions. These include the southeastern United States, with their hot, muggy summers, northern California with its long, dry summers, and the rainy Pacific Northwest. What they have in common are relatively mild winter temperatures.

- ✔ **Mid-latitude severe-winter climates:** These climates are controlled by conditions that develop over large land masses, such as North America's northern interior. As you might expect, they are generally north of the mild-winter climates. Their summers can be hot and humid or cool. Another variation is a "subpolar" climate of short, cool summers such as extends across much of Canada and Alaska. What they all have in common, as you might have guessed, is severe winters.

- ✔ **Polar climates:** No surprises here. Winters are long and severe. Even though summer days are long, daylight lasts many hours, the Sun never gets very high above the horizon, and so its light never really warms up the land. Polar climates extend across the Arctic, northern Canada, Alaska, Asia, and coastal Greenland. In the Southern Hemisphere, only the continent of Antarctica has a polar climate.

What Makes Climates Different?

Maybe you've already suspected this: The world's climate patterns are the result of persistently uneven ways that the Sun's heat energy and the sky's moisture are distributed around the globe. Climates have a number of main features and forces that shape them — the intensity of sunshine, the uneven arrangement of land and water, prevailing winds and ocean currents, persistent high and low pressure areas, mountain barriers, and elevation.

✔ **Sunshine:** The difference between the Tropics and the polar climates makes one thing pretty obvious: The longer a place gets intense sunshine, the warmer its climate. But the situation in the Tropics sets in motion a not-so-obvious process that affects climate everywhere. Powerful storms around the Equator send toward the poles a series of circulation cells that create bands of heat and cold and wetness and dryness around the world. (In Chapter 4, you get a good look at these cells.)

✔ **Land and water:** It takes large land masses like the continents of Africa and North America and Eurasia to create extremely hot and cold temperatures. Oceans just don't act that way. Water absorbs the Sun's heat and holds onto it a lot longer. This is what climate scientists are talking about when they say the ocean has a long "thermal memory." In the tropical ocean, temperatures are warm, but they vary only a few degrees all year long. Land is a different story. Sunlight bounces right off of it, warming up the air above it. And when it gets dark, the heat absorbed by the land quickly radiates away. There seems to be almost no limit to the range of its extremes. The world heat record is 136 degrees in El Azizia, Libya, while the North American record is 134 degrees in Death Valley, California. On the cold side, the world record is 129 degrees below zero in Vostok, Antarctica, while the North American record is -81 in Snag, Yukon. The U.S. cold record, outside of Alaska, is -70 in Rogers Pass, Montana.

✔ **Prevailing winds:** Around the world, two sets of winds dominate climate. Near the Equator, *easterly winds,* which merchant sailors named the tradewinds, flow from east to west. At the mid-latitudes, the big storms of winter flow from west to east in the prevailing *westerlies.*

✔ **Ocean currents:** Big ocean currents deliver cool and warm water over great distances, and the warming and cooling affects climates in big ways. Off the U.S. West Coast, for example, the southbound California Current keeps the climate of California cooler than it would be without it. And in the Atlantic, the Gulf Stream bathes the coast of the Southeast U.S. in warm tropical waters and then heads northwestward all the way to England and northern Europe. If you take a look on a map or a globe at England and northern Europe and notice how far north they are, you might think winters there must be colder than the dickens. But because of this long conveyor-belt of warm water of the Gulf Stream, these regions enjoy warmer climates.

✔ **Highs and lows:** The great thunderstorms in the Tropics loft moist air high into the atmosphere that eventually comes down dry in fairly predictable places. This descending air creates high pressure over the Southwestern deserts of the U.S., making its climate hot and dry. Where it occurs over the northern Pacific Ocean, you find the Pacific High that dominates California's summer climate. The clockwise circulation is stable, relatively cool and dry over the West Coast, making for long, dry summers over the region. Off the Atlantic Coast, another persistent high pressure area — the *Bermuda High* — has the opposite effect. The Southeastern U.S. is on the other side of the high pressure. In the eastern two-thirds of the U.S., summers often are hot and muggy because winds circulating clockwise around the Bermuda High are less stable and carrying warm tropical moist air up from the Gulf of Mexico.

✔ **Mountains:** Air flowing into the side of a mountain has the same effect as water in a river flowing up against a large rock. It rises up and over the mountain. As it does this, the rising air cools, causing clouds to form. Often it snows or rains. On the other side of the mountain, something very different happens. The air that has been dried out on the way up one side now falls back down the slope and warms up again. The climate on the side where the wind comes often is cool and wet. On the other side, in what is called the *rain shadow,* the climate is dry.

✔ **Elevation:** The distance above sea level of a particular place affects its climate in two ways. During daylight, a town on a mountainside 1,000 feet above sea level won't get as warm as a town down in the valley. Air cools as it rises, and the air over the mountain town has an altitude of 1,000 feet. It's not as cool as the air 1,000 feet above the floor of the valley, mind you, because it's getting the sunlight radiated off of the mountain. But it's still not as warm as the valley town. This relatively warm 1,000-foot air is fairly unstable, a condition that can lead to the frequent formation over mountains of *cumulus* and even *cumulonimbus* clouds, which Chapter 5 describes. At night, temperatures cool off more quickly in the mountain town than the valley town because the air has less heat-absorbing atmosphere above it: 1,000 feet-worth, to be exact.

As Chapter 4 describes, the warm air that rose as a valley breeze up the side of the mountain during the day more quickly chills at night and slides back down into the valley as a mountain breeze.

Climate and the Seasons

Remember last winter? Was it warmer than normal, or colder? Was it wetter or drier than most winters in your area? People at the Go Figure Academy of Sciences who work in the climate branch of weather science can tell you all about last winter and how it was different than average winters in every part of the world. But keeping track of such things isn't all they do. Climate science has a forward look to it now. Researchers are beginning to figure out *why* winters are different from one another. More interesting still, climatologists are even beginning to offer their own kinds of predictions. While a weather forecaster tries to tell you what tomorrow will be like in your city, a climatologist might try to tell you what next winter will be like in your state.

Shifting climate conditions have powerful impacts around the world. They affect where storms come from and where they go. They determine whether the jet stream — and the storm track — loops around the world like a piece of wet spaghetti or travels straight like a freight train. They affect the number of tropical storms and hurricanes that form every summer and autumn in the Pacific and Atlantic oceans. Terrible floods are caused by a season of torrential rains falling on an area that is not prepared for it. And droughts and famines can result from climate conditions that steer patterns of precipitation away from areas accustomed to rainfall. Historians now realize that shifting climate conditions have contributed to the rise and fall of whole civilizations over the centuries.

A climate of mystery

But wait . . . just a minute. Come to think of it, why should one winter be any different than any other? (Whoa! *Then* what would my neighbor and I talk about?) Really, if the seasons are shaped by the tilt of the Earth as it rotates around the Sun (see Chapter 2), how come this good, ol' reliable astronomical arrangement doesn't make the seasons as regular as clockwork? If all that matters is the Sun's position in the sky, one winter ought to be pretty much like any other. (Weather would be pretty boring in that case, and you would be reading about something else.)

Instead, of course, the seasons are full of surprises. Some winters are cold and snowy or rainy, full of storms and hardship, while others are so warm and dry they hardly feel like winters at all. What makes a hard winter hard, or an easy one easy? What's going on here?

Here's what your climate people have been finding out: Just like weather, there are things happening on Earth that cause climate to change. There is such a thing as *natural climate variation*. It just happens at different time scales than weather. Instead of changing from day to day, climate changes from year to year or decade to decade or time scales that are even longer. Also, climate shifts are more subtle, not so easy to notice. They don't come through like a storm. It takes several storms to recognize the signs of an especially hard winter, for example, and some weeks have to go by before you decide that an especially mild winter season is at hand. Natural climate variations over long time scales are even harder to pick up.

Could it be sunspots?

Could it be sunspots that cause the variations in the same season? Is that why one winter is so different from another? Sunspots would be a pretty well-educated guess. After all, if the changing position of the Sun in the sky accounts for the differences between winter and spring, summer and fall, perhaps the 11-year cycle of highs and lows in sunspot activity on your star changes the seasons themselves. A perfectly plausible explanation.

Experts — astronomers, weather scientists, mathematicians, the best in the business —
have been looking into this idea for a long time. They're still working on it. As it happens, the Sun's radiation difference from high to low is only about one-tenth of 1 percent, but it may be that Earth's atmosphere magnifies these effects. If this is true, it may be that these cycles affect the climate from one decade to the next.

For the differences you see from one winter to the next, apparently the short answer is No, it is not sunspots. For the longer timescales of climate, researchers still aren't sure.

"It's the ocean, stupid"

(Question: Who was it that first told a presidential candidate: "It's the economy, stupid." Answer: I have no idea.)

Naturally, you look to the sky when you think about weather and storms. But the sky isn't going to tell you much about the seasons. Things are not so obvious when it comes to climate questions like the year-to-year differences in the seasons.

Here is a big idea in weather science: The atmosphere responds in a big way to what's going on underneath it. When it flows over ice sheets and snow fields, for example, it gets the chills. When it flows over the continents, it picks up some of the heat or some of the cold of the land.

But remember this large fact about your favorite planet: Earth is covered with water more than anything else — 70.8 percent, to be exact — and the ocean is where a lot of the weather and climate action is. In fact, the sea and the sky influence each other so strongly that weather scientists have come to refer to them as parts of a single coupled system. Get this idea down and mention it as often as you can. "Well, of course," you say, "it's a *coupled system*." And there you are, right at the cutting edge of weather science!

Working with this idea, climate researchers have been teaming up with oceanographers in recent years to see how all this water on Earth affects seasonal differences and other weather trends. They don't have all the answers yet, but they're finding out some very interesting things. That's what names like El Niño and La Niña are all about. They are labels that climate scientists have given to certain sets of *coupled* ocean-atmosphere conditions that cause seasonal and longer term changes in weather patterns around the world.

Pacific Body Parts

Want to know why Seattle has a reputation for cloudy skies and rain? Or why sunny Southern California is sunny?

The answers to these climate questions have a lot to do with what is going on thousands of miles away in the Pacific Ocean. The big Pacific storms of the mid-latitudes are usually pushed into the Pacific Northwest — and normally not into the Southwest — along a path traveled by the jet stream.

The oceans of the world have features such as prevailing currents of warm and cool water and layers of different temperature and *salinity,* or salt content, that are important to weather and climate thousands of miles away. The Pacific Ocean has a certain look to it, a certain arrangement of features like parts of a body, that is familiar to someone who knows what they are looking for and has all the gear to see it. All it takes is a few satellites and ships and submarines and balloons and buoys, what-have-you — and the know-how of modern climate science. Here's what it looks like.

The Warm Pool

Across the Pacific Ocean from South America to Asia is a big layer of cold, deep ocean water and a thinner top layer of warmer water. Normally, the warm top layer is shaped like a wedge — thinner on the side near South America and thicker over against Asia. This is because the tradewinds are blowing, constantly dragging warm surface water with them from east to west. (For more on tradewinds, see Chapter 4.) This piled-up warm water makes a big warm pool (technically known as the *Warm Pool*). Above this pool form big, drenching thunderstorms that regularly irrigate the great rainforests and jungles of Borneo and south Asia. Come to think of it, those rainforests and jungles wouldn't be there if it weren't for those storms.

Anyway, this pool and these thunderstorms are the business end of a big circulation loop that is formed between the ocean and the atmosphere. Some moisture from these storms finds its way into the Pacific jet stream, and so a lot of it ends up in the streets of Seattle. In a way, the Warm Pool acts like a boiler-room, pumping big flows of warm air and wet clouds high into the sky.

Could it be volcanoes?

Could volcano eruptions cause the climate to change? The answer is Yes.

Eruptions of large volcanoes pump enormous clouds of stuff high into the upper portion of the atmosphere known as the *stratosphere.* Some of the very big ones spew out tons and tons of sulfur gases and tiny ash particles, and researchers have come to realize that these eruptions can affect weather around the world for years afterward.

While the heavier ash and dust particles settle out of the stratosphere over a period of months, the sulfur gases combine with water vapor in the sunlight and form a reflective veil of sulfuric acid particles. These particles are bright and shiny and cause sunlight to bounce back out into space. The result is less sunlight reaching Earth and cooler temperatures around the world.

So it was that the year 1816, the infamous "year without a summer," followed the massive eruption in 1815 of Mount Tambora in Indonesia. More recently, the eruption of El Chichon in Mexico in 1982 and Mount Pinatubo in the Philippines in 1991 caused temperatures to drop around the world.

The Cold Tongue

On the eastern side of the tropical Pacific, the ocean up against the shores of South America near the Equator has a very different look to it than it does on the opposite shores near Asia. Where there is a warm pool of water piled up in the western Pacific, cool water usually stretches out from the shores of the eastern Pacific. This is part of an ocean-atmosphere feedback loop called the *Walker Circulation.*

The warm air that flows into the atmosphere above the Warm Pool in the west travels east, giving up heat and moisture as it goes, and when it falls back to Earth around South America, it is cool and dry. Here it gets caught up in the tradewind flow, the westbound surface winds and again it heads back toward Asia, along the surface of the sea. As these winds blow across the surface, the sun-warmed water on top of the sea goes along for the ride. In the ocean, the removal of this warm surface water has the effect of causing colder water from the depths to rise up and take its place. This water forms a long *cold tongue* that sticks out along the Equator from South America toward the warm pool. (Technical name: the Cold Tongue.)

Notice this close link between the sea and the sky, what scientists call a *feedback loop.* The atmosphere's constant winds cause the Warm Pool to form across the ocean. The ocean's warm pool causes these big storms to rise high into the atmosphere. The storms drive a giant circulation pattern that causes the winds to blow. So around and around it goes, like a dog chasing its tail. The people at the Go Figure Academy of Sciences report that sometimes the atmosphere seems to be in charge, and sometimes it looks like it's the ocean. So they can't tell which is the dog and which is the tail.

El Niño, His Cool Sister, and Their Kissing Cousins

Just when you think you've got the seasons in your area all figured out, along comes a winter that is nothing like what you expect. Around the country, around the world, everything seems to be upside down. Places like Seattle that are accustomed to cool temperatures and a lot of rain instead get warm, dry weather. Places like sunny southern California, the Southwest desert, and even the southeastern United States get cold temperatures and heavy rain. In the usually cold Northeast, it doesn't even feel like winter. And it snows in weird places like in Guadalajara, Mexico.

That's almost a sure sign of El Niño — above-average Pacific Ocean sea surface temperatures and atmospheric conditions that change the places where storms go.

Cold temperatures and rain befall the usually dry deserts, and warm dryness comes to the usually cool forests of North America, South America, and much of the rest of the world. The important thing is not that there is more hard winter weather around the world during El Niño, although during a big El Niño, a grande El Niño, there certainly seem to be more storms. El Niño's "Bad Boy" reputation as a natural disaster comes not from extra storminess, but from the fact that it puts storms in places that can't handle them so well.

Is El Niño a bad boy?

Is El Niño really bad news? It all depends on where you live. A powerful El Niño is terrible news to people in Ecuador and Peru, for example, who often face deadly floods in their steep desert terrain. It's not much good to southern California either, and for the same reason — the desert soil can't absorb much water, so the rains bring on floods and mud slides. But the folks in Seattle and western Canada don't mind a mild winter once in a while. In the northeastern United States, often El Niño means milder winter temperatures and less snow than usual — less hardship and a welcome savings on winter heating bills. And for people across the Gulf States and the Eastern Seaboard, El Niño does one really good thing in the Atlantic Ocean in the summer and fall. It generates powerful high-altitude winds that cut the tops off of tropical storms before they can grow into hurricanes.

Good or bad, scientists say that, aside from the seasons themselves, El Niño is the most powerful climate force on Earth. So this is a big part of the big answer to the little question: Why is one winter different from another? It's not the only reason, of course. But if one winter is *very* different from normal — almost the opposite of what you would expect — take a look out in the tropical Pacific Ocean.

El Who?

A few centuries ago, fishermen in Peru noticed that occasionally a warm ocean current would begin flowing south off their coast. Usually they get a cool current flowing north. This warm southerly current would wreck their fishing. They noticed that it always seemed to arrive around Christmas time. So they named it after the Christ Child, El Niño.

It wasn't until the 1970s that researchers began to realize that El Niño wasn't just a local shift in ocean currents off South America. It is part of something much bigger. Now the name El Niño refers to a whole set of ocean and atmospheric conditions in the tropical Pacific Ocean that affect seasonal weather patterns over much of the globe.

El Niño has been coming and going for thousands of years. Like much of weather, it seems to be part of Earth's way of blowing off extra heat that builds up unevenly in the Tropics.

Experts are not really sure why it happens when it does or what causes it to start or to stop. But they are getting very good at detecting El Niño conditions as they develop in the ocean and at seeing the effect it has on weather around the world.

Off the coast of Peru, meanwhile, when El Niño comes, fishing isn't any better than it ever was. Starting in the far western Pacific usually, powerful eastbound winds suddenly start pushing the warm pool back toward the central Pacific. This causes the big thunderstorms to follow the warm water out toward the center of the ocean. If it goes far enough — for whatever reason — the big circulation loop is thrown out of whack, and the tradewinds become weak. When this happens, the warm water acts like an outfielder chasing a long fly ball. Back back back it goes — it's out of here! You've got El Niño on your hands!

The El Niño look

If El Niño is out there, as it is in Figure 6-1, the Pacific Ocean has a very different look. Now the Warm Pool and the big thunderstorms (see previous section) have moved. Instead of up against south Asia, they are spread out into the middle of the Pacific or even far over on the other side of the ocean, up against the coast of South America. All kinds of big changes are taking place. The tradewinds have completely pooped out and may even be blowing in the opposite direction. The Cold Tongue is a goner, and in its place is extra-warm water. Take a look at the jet stream, the storm track. Instead of looping around like it normally does, it is so strong that it carries storms straight over the Pacific Ocean and across the Southwest and the southern United States.

These important season-changing conditions — the Warm Pool and the big thunderstorms and the winds — seem to swing back and forth from one extreme to another across the tropical Pacific Ocean. On average, every three to seven years El Niño seems to come along. But "average" and "normal" are not very good words to use with things like El Niño. Like a lot of things about weather, "average" and "normal" don't seem to come around very often. The swings from especially warm to especially cool Pacific Ocean conditions are not regular, and their pattern is not reliable.

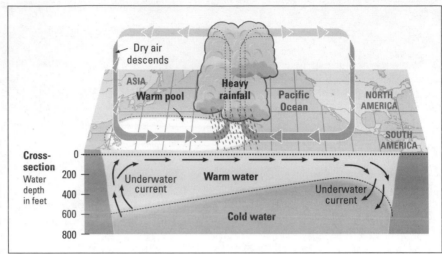

Figure 6-1:
Features of
El Niño
across the
tropical
Pacific.

Some climate experts at the Go Figure Academy of Sciences and elsewhere have been trying to make a computer model that acts like the Pacific Ocean, swinging between warm and cool, so that they can predict when El Niño will show his face. But they're not very good at it yet. On the other hand, scientists *have* become very good at using new tools like satellites and buoys moored across the ocean to detect changes in the temperature of the water and in the winds. So once El Niño conditions begin to take shape, they can see them sooner. If you live in a region that feels El Niño's effects, this means that some years climate experts may be able to warn you months in advance about what kind of winter to expect. Figure 6-2 shows the patterns of global weather that are linked to a powerful El Niño.

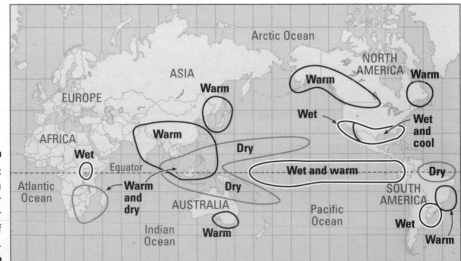

Figure 6-2:
Common
winter
weather
impacts of
El Niño.

The 1982–'83 "El Niño of the Century"

When a powerful El Niño came along in 1972, it caused drought in some agricultural areas that led to food shortages in parts of the world. But the first really big El Niño that came to everybody's attention was in 1982–'83. It brought torrential rains that caused terrible floods. In other places, it brought drought that led to enormous wildfires.

Researchers had been working on El Niño for several years by then, but the 1982–'83 El Niño really caught them by surprise. They didn't expect it, and even when the Pacific's surface was warming dramatically, the scientists couldn't agree that El Niño was on the way. Now they know that all El Niños are different from the "standard model" El Niño in the way they begin or end or their impacts on weather around the world. They all surprise researchers in one way or another.

This great El Niño of 1982–'83 caused about 2,000 deaths around the world, and its fires and floods caused damage losses estimated at $13 billion at the time.

Because nobody had seen anything like it, this one became known as "The El Niño of the Century."

La Niña, the contrary sister

El Niño causes some serious weather problems around the world, no doubt about it, but La Niña, which Figure 6-3 illustrates, is not exactly a sweet little thing either. In places like the northeastern United States that are accustomed to cold and snowy winters, La Niña often makes for especially hard winters. In the rainy Pacific Northwest, La Niña winters seem to bring even more rain and snow than usual.

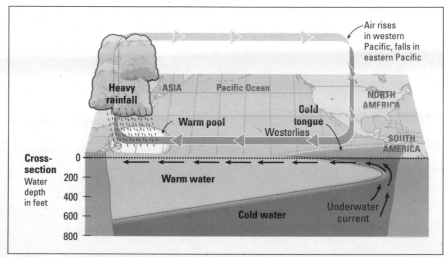

Figure 6-3: Features of La Niña across the tropical Pacific.

The 1997–'98 "El Niño of the Century"

In 1997, climate researchers were surprised again by the beginning of another big El Niño — just 15 years after "The El Niño of the Century." They had learned a lot since the El Niño of 1982–'83, and several fancy computer models at climate research centers around the world had predicted that an El Niño was on the way. But nobody saw a big one coming.

Still, there was an important difference. Now they could detect ocean temperature changes much more accurately, and they could measure these changes as they happened, in "real time." This meant that — for the first time — climate specialists at the National Weather Service were able to warn the public *months in advance* that a powerful El Niño was going to dominate the winter of 1997–'98.

For all the advance warning, still El Niño devastated much of the world with floods and fires. In the end, it was blamed for 23,000 deaths and $32 billion in damages worldwide.

In the media, this El Niño earned a familiar name: "The Climate Event of the Century." Go figure. . . .

Across the desert Southwest, often the season is even drier than normal. Tornadoes seem especially numerous during springs and summers of La Niña, and the Atlantic hurricane season can be especially long and dangerous. In 1999, for example, while La Niña conditions prevailed in the tropical Pacific Ocean, 12 tropical storms grew big enough to earn names, eight of them became hurricanes, and five became intense hurricanes.

Here is a rule that is not always exactly true, but still is useful to compare the impacts of El Niño and his contrary sister. Where El Niño is warm, La Niña is cool. Where El Niño is wet, La Niña is dry. While El Niño conditions and their seasonal impacts look very different from normal, La Niña conditions often bring winters that are typical — only more so. There's something else to keep in mind: El Niño and La Niña *tend* to make seasonal conditions one way or another, but every El Niño and La Niña is different.

The La Niña look

Like a lot of brothers and sisters, El Niño and La Niña don't get along together. In the tropical Pacific Ocean, they are opposites. If you have one, you can't have the other. In the ocean, the uppermost layer of warm water — which flattens out during El Niño — now has an especially sharp wedge-shape to itself during La Niña. It is thick up against South Asia, where the Warm Pool (see the section earlier in this chapter) is especially deep. Above the ocean, the tradewinds are especially strong. And on the opposite side of the ocean, up against South America, because of the strong off-shore winds, the warm surface layer disappears altogether. The deeper cold water is flowing directly up to the surface for a great distance along the Equator. This means that the Cold Tongue is sticking out far into the sea.

La Niña impacts on the world's weather are less predictable than the effects wrought by El Niño. This is mainly because of the big differences in the jet stream and the storm track (see Chapter 4). El Niño causes the Pacific storm track to become stronger, to drop farther south than usual, and to straighten out like a necklace of weather extending more-or-less straight across the ocean. The La Niña storm track is weaker and loopy and irregular, like a piece of wet and wiggly spaghetti, more changeable — so the behavior and direction of the storms it carries are more difficult to accurately forecast (see Figure 6-4).

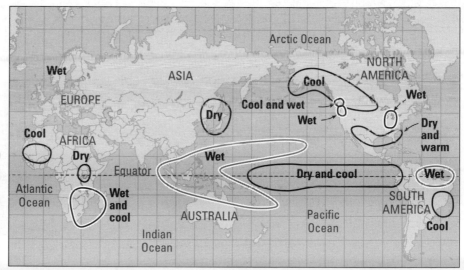

Figure 6-4: Winter weather impacts linked to La Niña.

Climate's kissing cousins

El Niño and La Niña are just the biggest and most obvious of an assortment of large-scale and long-term climate patterns. These large swayings in the atmosphere and sloshings in the ocean, warmings, and coolings all seem to be linked together in complicated ways that climate researchers are not yet very sure about. As they find out more about these big patterns, they hope to be able to more accurately predict their different impacts on weather and seasonal trends around the world.

This is the cutting edge of climate science, and these kissing cousins are the kinds of things that the people who are trying to make long-range climate predictions and weather forecasts are working on.

Scientists use a big, unfriendly word to describe these patterns they see in the atmosphere and the ocean. The unfortunate word is *oscillation*. An oscillation is a cycle that has the shape of a wave, or see-saw pattern. Peaks and valleys. Highs and lows. Hots and colds. Ups and downs. These are oscillations. A teeter-totter oscillates — one end is up when the other end is down.

Like El Niño and La Niña, these oscillations also act like "master switches" on seasonal weather patterns. Here are the two most prominent that researchers are working on:

- **Pacific Decadal Oscillation (PDO):** In periods of two or three decades at a time, researchers say, the Pacific Ocean north of the Tropics goes through cold and warm stages. Sometimes it's cold in the western and central north Pacific and warm off the coast of North America. And sometimes the reverse is true. These two phases have some of the same types of effects on North America's weather as El Niño and La Niña. This longer term north Pacific pattern works together with El Niño and La Niña, strengthening or weakening their impacts according to the phase of the Pacific Decadal Oscillation, or PDO.

- **Arctic Oscillation (AO):** Recent research indicates that the track of winter storms across much of North America, Europe, Russia, and northern Asia is affected by a large pattern of air pressure differences between the polar cap and a ring around the Arctic Circle. These pressure differences affect the winds. In one phase, the winds steer storms over Alaska and Scandinavia, and in another, the storm track drops down across California and Spain. (So it tells you when the rain in Spain is, mainly. . . .)

Climates of the Past

The biggest problem with studying the important changes in the climate is that the record of them is too short. How can you tell what to expect in the future if you cannot see into the past? The answer is you can't. Most of even the longest weather records kept with accurate measurement instruments hardly go back more than a century. A hundred years may sound like a lot of weather, but it's not nearly enough of a climate record to figure out what's going on.

For example, some climate cycles that affect the tracks of storms over North America may change only once every two or three decades or even less frequently than that (see previous section). Also, the frequency of Atlantic hurricanes may go through cycles lasting 25 years or more (see Chapter 11). Accurately identifying changes in climate cycles such as these can affect what kind of seasonal changes you and I might expect.

Also, climate scientists are on the hot seat, so to speak, to answer some other important questions being asked of them today. Are you and I facing a difficult future of increasing global warming caused by human emissions of industrial greenhouse gases? (This question is explored in detail in Chapter 13.) Or are the changes you see taking place in weather patterns all part of the natural ebb and flow of Earth's processes? Are you heading back into another "ice age?" For answers to questions such as these, researchers are looking as far as they can into the distant past.

Handicapped by a short instrument record, climate scientists who want to see where big, long-term weather trends are likely to go from here have had to use their ingenuity. To see further back beyond the last century or so, researchers have borrowed some methods and tools from the science of *geology,* the study of Earth's land masses. From their studies of such things as growth rings on very old trees and from layers in cores bored deep into polar ice sheets, scientists are able to estimate what the climate was like way back when. Because these estimates are indirect evidence rather than direct measurements, they are called *proxy* records.

Long warm ages

Did you know that for long periods of time the Earth's climate was a lot warmer than it is today? Some researchers estimate that as little as 10 percent of Earth's past has been ice ages. Others think it's longer than that. But still, researchers agree that for hundreds of millions of years it was between 14 degrees and 25 degrees warmer than now. For example, one of those warm periods was about 65 million years ago — about the time the dinosaurs became extinct.

Imagine such a place: no glaciers anywhere, and no ice caps at the poles. All of that water that is now locked up in glaciers and ice sheets would have been part of the oceans, so the oceans would have been bigger. And places along the shore that today are high and dry would have been under water, part of the floor of a bigger ocean. Giant, barren deserts like the Sahara at one time were lush with tropical vegetation and teeming with animals.

Between these warm periods, however, were other long stretches of much cooler temperatures. What has surprised many researchers is recent evidence that the switch from one climate phase to another — from ice age to warm period — occasionally has taken place fairly quickly, maybe in only a few years.

Short ice ages

According to the current ballpark estimates, Earth's long warm past has been interrupted by seven ice ages that have been relatively brief. While the warm periods lasted hundreds of millions of years, ice ages are measured in tens of millions of years.

About 700 million years ago was a big Ice Age, maybe the biggest of all. In fact, there is evidence that leads some researchers to think that during this ice age, ice covered the earth nearly everywhere, all year round. The entire planet may have been a big snowball.

Did you know that you and I are living in an ice age now? There are glaciers and ice sheets on the continents, particularly in the Northern Hemisphere, and ice caps cover the Arctic and Antarctica, the North and South poles. Climate scientists will tell you that the current ice age began gradually about 55 million years ago and so covers all of the time that there have been humans on Earth. It reached its maximum extent about 20,000 years ago, when its ice sheets, as much as two miles deep, had spread down over the northeastern United States and the Midwest. This ice age has been in retreat since then, but even during the last 15,000 years, when the current warming began, sharp climate changes have interrupted things.

Medieval warm period

During most of human history, the giant far northern continent called Greenland has been anything but green. It has been inhospitably cold and covered with giant ice sheets most of the time. About 1,100 years ago — the year 900 — climate in the North Atlantic suddenly became warmer.

Greenland got its name from this time, when the Vikings of Norway were powerful explorers. They had settled Iceland, and their North Atlantic adventures led them to discover the big green pastures of the giant continent to the west. The ambitious Norsemen settled Greenland. But about 1200, the climate began to grow cold again. Within about 200 years, the Viking settlers were stranded, and the Greenland colonies were lost.

Scientists are not really sure that this "Medieval Warm Period" deserves a global climate name of its own. There is evidence that what happened in the North Atlantic was regional change prompted to a shift in ocean currents. But there's no doubt about what happened next.

Climate detectives

It doesn't take a genius to figure out some climate evidence that geologists have dug up. The long scrapes left by a glacier passing over the surface of a big rock are pretty unmistakable. The material left over from such events as floods also can be found underground without too much trouble. But this kind of evidence is not good enough to construct a picture of what Earth's climate was really like at different times in the past. For that picture, they have had to employ some pretty shrewd detective work. Here are some of their favorite methods:

✔ **Tree rings:** In forested areas, researchers can read the annual growth rings of old trees and tell how climate changed over their lifetimes (see figure). Times of plentiful rain or snow are marked by rings that are thicker than those left by times of drought.

studying deep ice cores can read the thickness of these layers, and they can analyze the chemical composition of the dust deposits and tiny air bubbles trapped inside the ice and discover all kinds of things about ancient climate (see figure).

Courtesy of John Rhoades.

✔ **Sediments:** On the floor of the oceans and large lakes are deep accumulations of ancient plant and animal life and other debris that climate detectives find fascinating. From the chemical makeup of the shells left by long-gone tiny aquatic organisms, for example, they can tell how much oxygen or carbon dioxide was in the atmosphere at the time.

✔ **Pollen:** From the pollen of ancient plants preserved in lake sediments, researchers can figure out what types of vegetation grew in a place. With that information, they have a good idea what kind of climate prevailed when the pollen was part of a living plant.

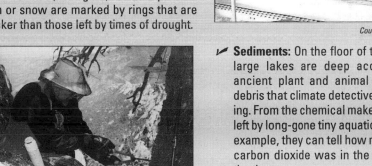

Courtesy of John Florence, University of Arizona.

✔ **Ice cores:** The seasons leave their own permanent signatures on the glaciers and ice sheets, which do not melt. Winter is white and thick, summer is thin and darkened by the accumulation of bits of stuff deposited by the atmosphere over months. Researchers

Little ice age

The period of warmth during the Middle Ages was soon followed by a cold spell that began about 1450 and lasted 400 years. It wasn't really a new ice age, but it sure felt like it to the people of western Europe and North America.

This was a time of long, severe winters and hardship and short and wet summers in Europe and North America. The latest studies indicate that the year 1601 was the coldest of the last 1,000 years, and the following century was the coldest of the millennium.

Agricultural production failed in many areas of Europe, leading to famine. In North America, the new colonies were hard hit. The English colony at Jamestown was almost abandoned after the terrible first winter in 1607. More than half the colonists starved or died from the cold.

A warming trend

Temperatures have been less cold and winters have been less extreme in the 1900s, but there is a definite climate trend around the world over the past century. As you may have guessed, it is getting warmer.

Annual surface temperatures around the world warmed 1.12 degrees from 1901 to 1997. The warmest years on record all occurred in the 1990s. The year 1998 was the warmest in recorded history. In fact, scientists say it may have been the warmest of the last 600 years.

What is causing this warming? Is it just part of Earth's natural climate variation? Or is it caused by human activity, by the so-called greenhouse effect that industrial gasses like carbon dioxide have on the atmosphere? I explore all these questions about global warming in Chapter 12.

Chapter 7

The Greatest Storms on Earth

*N*ot for nothing are they called the greatest storms on Earth. Few things in nature are more powerful than hurricanes. The giant tropical storms that come swirling out of the warm waters near the Equator like enormous coiled springs wreak death and destruction across the world every year. In an average year, 85 tropical storms form around the world, including 45 which will reach the wind intensity of at least 74 miles per hour, and so be classified as hurricanes.

This chapter looks at hurricanes from the inside out. Where do they breed? What makes one hurricane season busy and another season quiet? How well can weather forecasters predict where these destructive giants will make landfall? In this chapter are sections that focus on answers to these questions and more.

On average, the National Weather Service reports, ten tropical storms form every year in waters affecting the United States — the Atlantic Ocean, the Gulf of Mexico, and the Caribbean Sea — and six of them typically become hurricanes. In the United States, hurricane damage currently costs an average of $6.2 billion a year.

In an average three-year period, about five hurricanes strike the U.S. coastline anywhere from Texas to Maine, killing from 50 to 100 people. Two of these five are major hurricanes with winds greater than 110 miles per hour.

In 1999, more than 11,000 people died during a single storm in Honduras and Nicaragua.

Breeding Grounds

A *hurricane* is a tropical storm run amok — a rotating mass of thunderstorms that has become highly organized into circular cells that are ventilated by bands of roaring winds.

In the North Atlantic Ocean and the eastern North Pacific Ocean, they go by the name *hurricanes*. In the western North Pacific, they are called *typhoons*, and in the Indian Ocean and everywhere south of the Equator they are known as *tropical cyclones*. By whatever name, they are terrible storms that can roam across thousands of miles of ocean and last anywhere from just a few hours to as long as a month.

In the western Pacific, the region known as *Typhoon Alley*, they can strike just about any time of year. In the North Atlantic, the hurricane season that threatens the Gulf Coast and the East Coast of the United States begins in June and continues through the end of October. This is the time of year when the water in the Atlantic Ocean north of the Equator is at its warmest. By October, the water in the region is beginning to feel the effects of autumn as the Sun's direct rays shift toward the ocean south of the Equator. High season for Southern Hemisphere cyclones is December through March.

Storms that form in the Tropics are different from the big frontal mid-latitude storms that more commonly sweep across the countryside where you and I live. The big storms of winter may be seen as creatures of the sky, powered by temperature contrasts and the energy from the winds they create. Tropical storms seem more like ocean critters that form in a uniform environment and are fueled by energy from the latent heat of their condensation.

Under the microscope

Hurricanes are the subject of a lot of research by weather scientists in the United States and elsewhere. They are a big problem and a major focus of the U.S. Weather Research Program.

Under the five-year program, federal agencies are teaming up with university researchers and scientists at other institutions to tackle some top-priority weather forecasting goals. First on the list is a set of ambitious goals for improving hurricane predictions.

Scientists want to be able to better forecast where the hurricane will hit land, how much time people have to evacuate before winds get too strong, and just how strong the hurricane is going to be. They want to increase warning lead-time from two days to four days, to narrow the length of warned coastline from nearly 400 miles down to 200 miles, and to cut in half their errors for forecasting a storm's strength.

Other parts of the U.S. Weather Research Program focus on improving forecasts of winter storms and spring and summer rains that bring heavy precipitation and the threats of flooding. Scientists also are looking more closely at the sources of the storms that sweep into the U.S. West Coast from the Pacific Ocean.

Typhoon Alley

An average of 31 tropical storms roam the western North Pacific every year. The Island of Guam, a U.S. territory, has been hit by 16 typhoons since 1970. And since 1960, four typhoons have devastated the island. In December 1997, Super Typhoon Paka smashed into the island with winds of 150 miles per hour and gusts up to 185 mph.

Why does Typhoon Alley get so many typhoons? And how come they can take shape in the western North Pacific all year long? Like a lot of mysteries about hurricanes, the answers to these questions are found in the ocean. The key is the *Warm Pool* of ocean water that Chapter 6 describes. All year long, the tradewinds and the ocean currents are pushing surface water warmed by the tropical Sun to the far western side of the North Pacific. Hurricane seasons come and go in other regions, but the water of the Warm Pool always is warm enough to hatch a hurricane.

Coastal communities around the world are no longer surprised by the arrival of hurricanes as they were many years ago. Before regular monitoring flights by airplanes began in the 1940s, the storms on the high seas would come and go without anybody knowing it except a passing ship now and then. The worst natural disaster in United State's history was a "surprise" hurricane. It came ashore in September 1900 and swamped Galveston, Texas, with its storm surge, killing more than 8,000 people.

Now forecasters can see hurricanes coming several days in advance. With aircraft roaming the skies and satellites circling the planet, there's no place for a hurricane to hide. Equipped with powerful computers and monitoring systems and instruments, weather scientists and forecasters continue to make progress which no doubt has saved thousands of lives in recent years.

Mysteries

Advances are being made, but still, important things about hurricanes are not well understood. Forecasters and researchers can see them take shape in the ocean, but they cannot very accurately predict their behavior. Why does one tropical storm build into a hurricane while another just peters out? Weather scientists don't know exactly. They also cannot precisely predict where or when a hurricane will become stronger or bigger, or when its forward motion will speed up or slow down or change direction.

For weather forecasters, an approaching hurricane is a very big and complicated problem. From a single storm, thousands if not millions of lives can be in jeopardy, and several hours are needed to safely evacuate large numbers of people from coastal communities that appear to be in harm's way. And yet, as the section of this chapter, "Coming Ashore . . . But Where?" points out, forecasters can't be sure just where or when an approaching hurricane will come ashore. And when one does, they cannot very accurately forecast how high the big tidal sea bulge known as the *storm surge* will be when it slams into coastal areas. They can't be sure how much rain will fall when the hurricane moves inland or how many tornadoes the hurricane will spawn that might rake the countryside. From one year to the next, they can tell you that one Atlantic season is *likely* to see more hurricanes than another. This is helpful information — even if they can't really tell you how many to expect.

These are mysteries that hurricane forecasters throughout the world face all of the time. Figure 7-1 gives you an idea how big the problem is.

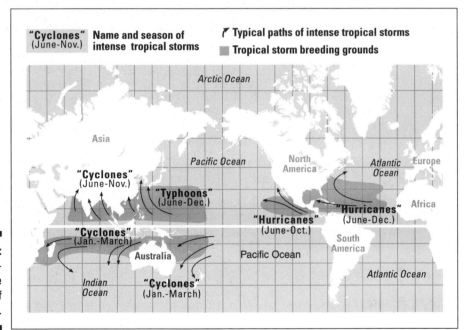

Figure 7-1:
Hurricane-
prone
regions of
the world.

Birth of a Hurricane

When conditions are right, a thunderstorm cloud takes shape very easily in the breeding grounds for hurricanes. In the Tropics, conditions are often just right. The Tropics are like a big room full of young people listening to loud music. Before long, there is dancing. The ocean's surface is 80 degrees or warmer, ready to evaporate into the air at the slightest chance. Winds are light and fairly evenly spread out in the sky, so air can rise high. And there is just enough disturbance to get that warm, moist air mixing upward, its water vapor cooling and condensing into water droplets — a cloud. This kind of mixing happens all the time along a band near the Equator where southward blowing tradewinds and northward blowing tradewinds come together and make a fold in the atmosphere.

Weather scientists have a five-dollar word for this thunderstorm belt — *Intertropical Convergence Zone*. Also, in the Atlantic Ocean north of the Equator, a jet of air flowing with the tradewinds from the east, out of North Africa, regularly develops wrinkles or waves in it. These *easterly waves* in the air flow are like musical beats in the Tropics — just the things to get thunderstorm clouds taking shape. In summer and autumn, dozens of these waves march westward across the tropical North Atlantic, although only a few grow to become hurricanes.

Disturbance to depression

A shapeless cluster of thunderstorm clouds known as a *tropical disturbance* may begin to dance around one another if conditions are right. The mass of cloudiness begins to get some spin to it. This is the force of the Earth's rotation at work — causing the clouds to twirl counterclockwise north of the Equator and clockwise south of the Equator. The winds are blowing in a circular formation, and at the center, an area of low air pressure is forming. The disturbance is not so shapeless anymore. Its thunderstorms are beginning to get organized. Winds around its weak low center are 23 to 39 miles per hour. It has become a *tropical depression*.

Storm to hurricane

What was once just a blob of tropical cloudiness on a weather satellite image continues to grow in size. Its overall shape is more circular, and with a series of satellite images, it's easy to see that the bands of clouds are rotating around a center. Winds around the center are blowing faster than 39 miles per hour. The tropical depression has become a *tropical storm* and has earned itself a name. Without ever becoming a hurricane, a tropical storm can dump huge amounts of rainfall as it plows inland and cause serious flooding.

A hurricane is a tropical storm, only more so. Air pressure has continued to drop at the surface of the sea in the center of the storm, and now the center often can clearly be seen in the satellite images. There is no mistaking the circular rotation of this creature. The whole storm now looks very much like a spinning pinwheel. The shapeless blob of a few days ago has grown into a highly organized weather machine. At the very center of the wheel of clouds is a dark spot, the *eye* of the hurricane, where often there are no clouds at all and winds are fairly calm. Around this clear dark hole in the storm is a sharp, steep *eye wall,* a high cliff of clouds. Around this central core, winds speeds have reached at least 74 miles per hour. And all around the hurricane, from the eye to the ragged outer edges are enormous *spiral rain bands* of cloudiness. Figure 7-2 shows what a hurricane looks like from the inside.

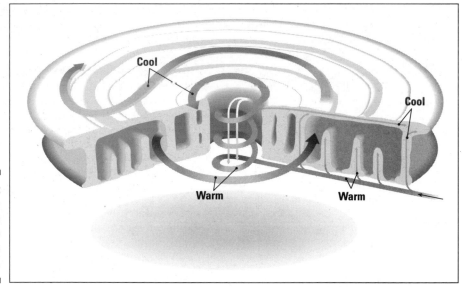

Figure 7-2:
What a hurricane looks like from the inside.

Bad big weapons

A hurricane throws everything at you in bad weather's big arsenal of weapons except the ones that come from extreme cold. Crashing lightning and thunder, it comes ashore with a tidal bulge and big waves that can swamp coastal communities. It whips inland with its famously powerful knockdown and blow-over winds, and with occasional tornadoes sweeping down from its clouds. Long after passing its peak, it pours flooding rains across the battered countryside far from the sea.

Storm surge

Along the coastline, the greatest threat to life and property is the *storm surge,* the large dome of water that is 50 miles to 100 miles wide that sweeps over the shoreline where the hurricane comes ashore. More than the winds, this is the hazard that wrecked Galveston, Texas, in 1900 when an estimated 8,000 people were killed (see sidebar). Danger is greatest if a hurricane makes landfall during high tide, especially in areas where there is a big difference between high tide and low tide. On top of a storm surge that can be more than 15 feet high are large wind-whipped waves of the storm. It is a lethal combination for harbors and seaside vacation resorts. Photographs in the color section earmarked Hurricanes illustrate the impact of storm surge.

Three things control storm surge: wind speed, water depth, and the very low air pressure in the eye of the storm that allows the sea to rise. The stronger the winds, the shallower the offshore water, the lower the pressure, the higher the storm surge. The Category 5 winds of Hurricane Camille, which struck Mississippi in 1969, produced a 25-foot surge.

The storm surge of hurricanes, illustrated in Figure 7-3, continues to kill thousands of people living in low-lying coastal areas around the world. In 1973, the floodwaters of a 23-foot storm surge from a cyclone drowned more than 300,000 people in Bangladesh. In 1991, a cyclone in the same area killed 140,000. But the death toll in the United States has dropped off dramatically in recent years. This is the big payoff for better forecasting and coordinated evacuation of the coastline. While property damages have shot upward, the buildings and houses are empty — lives lost now are far fewer. But experts agree that part of the statistics is a matter of dumb luck. Intense hurricanes making landfall in populated areas of the United States have been few and far between in recent decades.

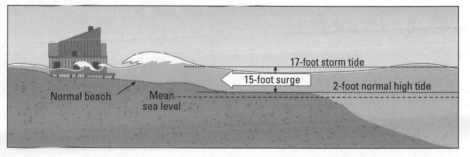

Figure 7-3: How storm surge takes shape along a coastline.

17-foot storm tide

15-foot surge

2-foot normal high tide

Normal beach

Mean sea level

Winds

Damaging winds spread the destructive force of a hurricane far and wide. They begin long before the storm surge or the eye of the hurricane come ashore and last long afterward. As Hurricane Hugo proved in 1989, hurricane winds can cause heavy damage very far inland. Its wind gusts near 100 miles per hour, which may have been from a tornado spawned by the hurricane, caused heavy damage 175 miles from the coast in Charlotte, North Carolina.

Winds of hurricane force, 74 miles per hour or greater, can extend outward from the eye for 25 miles for a small hurricane and to more than 150 miles for a large hurricane. Beyond these winds, powerful tropical storm-force winds, 35 to 73 miles per hour, can stretch out as far as 300 miles from the center of a large hurricane.

The *speed* of wind is fairly easy to get a grip on. All kinds of *anemometers* are around to do the job, as Chapter 1 describes, although it's hard to find one that will stand up to the winds of a hurricane. But the *power* of wind, the pushover power, is a tricky thing to measure. The wind pressure against a surface increases quicker than the wind speed. Category 4 hurricane winds blowing at 148 miles per hour aren't going to do *twice* the damage of a Category 1 74-mph winds, but more like *100 times* the damage.

For several years now, weather scientists have been taking a closer look into the wind flows of hurricanes to get a better grip on the damage they do. The wind motions turn out to be more complicated than the circulation pattern that most weather scientists had in mind not long ago. The late tornado expert Ted Fujita looked closely at the damage done in 1992 by Hurricane Andrew in southern Florida and found that the worst damage of Andrew was done by powerful whirlwinds inside the bigger hurricane circulation. At maybe 10 miles across, they are smaller than the hurricane, of course, and maybe ten times wider than a big tornado. Winds inside these small eddies that spin off the eye wall of the hurricane reach speeds of 200 miles per hour.

Tornadoes

The big thunderstorm clouds of hurricanes can threaten the countryside with tornadoes for days after the big storm comes ashore. While some tornadoes take shape near the eye wall, tornadoes are more likely to form at the fringes in the long spiral rain bands many miles from the center of the storm. The more intense the hurricane, the more likely it is to threaten tornadoes. Chapter 9 goes into the details about tornadoes.

Flooding rainfall

Hurricanes dump huge amounts of rain, often producing severe flooding. In fact, more than winds, inland flooding has been the main cause of hurricane-related deaths in the United States in the last 30 years. A typical hurricane brings 6 inches to 12 inches of rainfall. Heaviest rains come six hours before and six hours after a hurricane comes ashore, although it can continue to dump heavy rainfall long after it has begun to deteriorate. The biggest rains occur in hurricanes that move fairly slowly, less than ten miles per hour.

Hurricane Mitch

It started as a tropical depression on October 22, 1998, just another shapeless blob of clouds in the southern Caribbean Sea. Within a day, it was a tropical storm and given the name Mitch. Within four days, it was a Category 5 hurricane, a monster storm carrying winds of 180 miles per hour and gusts of more than 200 mph. For all that power, however, Hurricane Mitch will not be remembered for its winds.

The torrential rains of that great storm killed more than 11,000 people in Central America the last six days of that October. Just how big the toll of death will never be known. Years after the hurricane, thousands of people in Honduras and Nicaragua remained unaccounted for. And for years afterward, deep scars were everywhere across the mountainous countryside. In Honduras, most major bridges and secondary roads were washed out and many whole villages were swept from the mountainsides.

Estimates were that it would take 15 years to 20 years just to rebuild basic public services in the impoverished nation.

Total rainfall in some mountain areas totaled more than 6 feet — 75 inches — from Hurricane Mitch. The mountainsides literally dissolved under the torrents. Mitch stalled for days over the narrow strip of mountains. Its circulation pulled big tropical flows up from the tropical Pacific as well as the Caribbean. Even as its wind strength faded, the effect of the moist air flowing up both sides of the mountains created a continuous bomb of rainfall.

The deadliest Atlantic hurricane in more than 200 years finally drifted off into the Gulf of Mexico, regained strength, and on November 4 it pounded the Florida Keys with tropical storm winds and heavy rains.

Forecasters have a major problem on their hands predicting the amounts of rainfall during a hurricane. Its powerful winds are whipping over the landscape and changing the lower atmosphere's circulation constantly. Forecasters have a whole flock of circulating thunderstorms spiraling toward them in one big, thick band after another. With help from Doppler radar and monitoring aircraft, some overall estimates can be made. But like a lot of things about hurricanes, it's hard to be very right and easy to be very wrong. Some tropical storms can unload rainfall that is measured in feet rather than inches.

Signals of the Seasons

Weather scientists can't tell exactly how many hurricanes are going to form in the Atlantic basin, or anywhere else, one year to the next, and they can't tell in advance how many are going to make landfall, or where they will come ashore. What they can tell beforehand is this: It looks like its going to be a busy hurricane season, or it looks like its going to be quiet, or it looks about average. Forecasters even include specific numbers of hurricanes and tropical storms in their seasonal forecasts. These numbers are interesting to insurance companies, although my people at the Go Figure Academy of Sciences tell me they are not the kind of thing you would expect to take to the bank.

The problem with seasonal forecasts for Atlantic basin tropical storms and hurricanes is that the numbers are so small. When you're counting things on the fingers of one hand, it's hard to make statistical cases stick. You know, 20 is only about 5 percent more than 19, but 2 is 100 percent more than 1.

The long-term yearly averages for the Atlantic basin look like this: 9.3 tropical storms, 5.8 of them are hurricanes, and 2.2 of those are intense or major hurricanes. How about the really big question you want answered, the likelihood of a hurricane hitting home? About five hurricanes hit the U.S. coastline, somewhere from Texas to Maine, every three years. That's 1.67 per year.

From one year to the next, the numbers jump around like souped-up hotrods. It's hard to be right and easy to be wrong. Like: "I think there is going to be one major hurricane this year, but there could be two. And three, of course, is not out of the question." And then along comes a monster like an Andrew in 1992 or a Mitch in 1998, and suddenly the yearly averages look pretty darned beside the point anyway.

Researchers are getting a pretty good handle on how changes in some large-scale climate conditions affect seasonal activity of hurricanes in the Atlantic Ocean, the Caribbean, and the Gulf of Mexico. The five-dollar word for most of these is *oscillations,* which are see-saw patterns of temperatures or other features in the atmosphere or ocean.

William Gray is a famous hurricane researcher at Colorado State University who has pioneered seasonal forecasting and linked tropical storm activity to a variety of climate patterns. These include the amount of rainfall in western Africa, the direction of winds in the stratosphere over the Tropics, and other temperature and pressure oscillations.

The Pacific Ocean warming known as El Niño, which Chapter 6 lays out, really puts a damper on Atlantic hurricanes. Its warm waters in the eastern Pacific create strong westerly winds that rip the tops off of tropical depressions before they reach hurricane strength. La Niña, El Niño's sister, is the name for extra-cool Pacific Ocean surface temperatures and has the opposite effect, prompting more hurricanes than usual.

The bad news is, as the Chinese saying goes, you and I are living in interesting times. Dr. Gray has identified an important shift in the long-term trend of hurricane activity in the North Atlantic — and the trend is not good.

Hurricane modification

What if scientists could change hurricanes? Cut 'em off at the pass. Zap them in some way to stamp out their winds. Or steer them back out into the open ocean. The weather scientists who figured out how to make hurricanes less dangerous to people would be real heroes.

Off and on, scientists have been chasing this idea for many years. Here are a few things they have tried — or thought about:

✔ How about nuking them with a bomb? Just blow them to smithereens. This was a big idea in the 1950s, before the dangers of nuclear fallout were well understood. Nobody really tried it. Even if it worked perfectly, trading a hurricane for a radiation cloud circling Earth in the tradewinds does not sound like a bargain.

✔ How about *seeding* their clouds? Drain their clouds of water, and so weaken their winds. This strategy was tested for about 20 years, but finally — to make a long story short — scientists decided that it didn't work.

Now scientists are looking into the idea of spreading a chemical on the ocean that would make it hard for the hurricane to suck up the heat energy that it gets from the sea. But so far, nobody has come up with the right chemical.

In the meantime, it looks like the threat of hurricanes is something that people who live along the coastline from Texas to Maine are going to have to live with, hunker down against — or move away from.

Hurricane seasons seem to go along at one level of activity for 20 years or more, and then they shift into another gear. From 1900 to 1925, for example, hurricane seasons were pretty quiet, and then they picked up dramatically from the 1930s until the late 1960s. From 1970 to 1994 was another long quiet period. And now — you guessed it — the last years of the 1990s were the most active hurricane years on record. As the history of hurricanes has proven more than once, no matter how active the season, it only takes one storm to cause disaster.

Gray thinks these big shifts are caused by changes in the flow of water through the ocean, the big pattern known as the *Atlantic conveyor belt*. Changes in water temperatures and saltiness cause it to speed up and slow down over these long time periods, says Gray, and a faster-moving conveyor belt makes more active hurricane seasons and more major storms coming ashore along the U.S. East Coast.

Cruise'n for a Bruise'n

Hurricanes are weather's greatest storms and are the biggest single threat to life and property in the United States. Already, hurricane science has much improved in the last several years, but in two important ways, the threat has grown bigger rather than smaller.

✔ For one thing, a new long-term trend toward more hurricanes seems to have kicked in. The U.S. can expect more hurricanes each season, on average, than were likely a few years ago.

✔ For another thing, the country has made itself more vulnerable in recent years. People continue to move more and more of their homes and businesses and bodies into the paths of hurricanes.

Circumstances have reached this unfortunate point:

The time that it takes to evacuate some crowded coastal communities is longer than the time for which accurate hurricane forecasts can be made.

Although many hurricane scientists feel like they are fighting losing battles against these trends, they know that a lot of public interest is focused on the progress of their research. Forecasters and researchers are throwing everything they can into the job of following and figuring out these great storms — the best tools and some of the hardest thinking.

The following sections describe their most important tools.

Satellites

While the hurricanes are still far out at sea, forecasters rely on direct observations from ships and instruments on ocean buoys, but indirect measurements from satellites are their main tools. These satellites orbit 22,000 miles above the Equator at the same rate as the Earth's daily rotation, so always they hover over the same patch of the planet.

Their Big Picture images allow forecasters to track storms night and day all over the Atlantic basin. These images are the first important clues to the formation of these storms as well as their location, size, and intensity.

Aircraft

As the storms move closer to land, direct measurements are made by "Hurricane Hunter" airplanes.

Routinely, the U.S. Air Force Reserve uses a specially equipped flight of C-130 aircraft to monitor approaching hurricanes. Their pilots fly into the core of the storm. Specialists aboard these planes drop instrument packages through the storms that radio back accurate information about the location of its eye, the strength of its winds, the direction and pace of its progress, and other important features such as air pressure, temperature, and humidity.

The National Oceanic and Atmospheric Administration also has two special hurricane-probing planes, including a high altitude jet, that are designed for research.

Radar

As the storms get within about 200 miles of the coastline, land-based weather radar begins to give forecasters important indirect measurements. The new Doppler radars give forecasters valuable images that give details about the internal structure of the storm — its wind fields, and how they change as a hurricane approaches land.

Computer models

Forecasters are increasingly relying on computer models to help them forecast the intensity of the hurricane and its movement. Detailed information from all of the satellite sensors, and aircraft instruments, radars, ships, and buoys continuously pours into the computer models. The results of a whole group of forecasting models are compared with one another as hurricanes approach the coast.

The best of the bunch is a model operated by a supercomputer at the Geophysical Fluid Dynamics Laboratory in Princeton, New Jersey, which promises boosts in tricky track and intensity forecasts. This one most accurately mimics the way a hurricane works, experts say, and recent improvements are plugging in important information about ocean temperatures that are crucial to how a storm grows in strength.

At the National Hurricane Center in Miami, forecasters also use a special computer model devoted to forecasting a hurricane's storm surge.

Coming Ashore . . . But Where?

When a hurricane moves over land, usually it marks the beginning of the end for the big storm, because it is cutting itself off from its fuel supply — the warm water of the ocean. Unless it crosses back over warm ocean water at some point, eventually it will run out of gas.

Of course, this fact offers no comfort at all to the unlucky people whose homes and businesses lie in its path, or *track*. A hurricane is such a large storm, and the momentum of its circulation is so great, its winds and torrential rains can wreak havoc far inland from the seashore for several days. Long after it has weakened below the speed limits of a hurricane, a tropical storm still can swamp the countryside with flooding rains. Some dissipating hurricanes get caught up in other mid-latitude storm circulations that Chapter 8 describes.

The Big Picture

The Big Picture of the way hurricanes travel over long distances is not very complicated. They travel from east to west in the flow of the tradewinds both north and south of the Equator. And along the way they get caught up in other wind circulations and ocean currents that cause them to veer toward the poles and eventually slant back around toward the east. In the North Atlantic, for example, the tracks of tropical storms and hurricanes often curve northward in the wind flow around the Bermuda High and follow the warm waters of the Gulf Stream into the Gulf of Mexico or up the Atlantic coast. In the eastern North Pacific, the U.S. West Coast almost never encounters hurricanes. The tradewinds blow developing tropical storms away from the mainland, and those that do manage to turn northward run into the cold waters of the California Current that saps their strength.

But when a hurricane is on the way, people along the coast don't give a hoot about the Big Picture. Really knowing where and when a hurricane is going to move inland — that is terribly important information. First and foremost, when a hurricane is coming ashore, this is what you and I want to know:

- Is it going to hit me, or is it going to miss me?
- How much time have I got?

Answering these vital questions about the direction of its track and the pace of its advance are the kinds of things about hurricanes that keep weather forecasters awake at night. You might think that telling where such an enormously large storm is coming ashore would be a piece of cake. After all, its been out there chugging along in plain view for several days now, and there it is in the satellite photo, bigger than life, just off the coast. Like just about everything about hurricanes, however, predicting the exact track that it will take and exactly when it will move on shore is a lot more difficult than it looks.

Tracking the track

Hurricanes don't move at steady paces, and they don't move in straight lines. They stop and start and zig here and zag there all along the way. Figure 7-4 shows some sample hurricane tracks that give you a pretty good idea of the problem that forecasters face. It's a little like a long-distance airplane flight. A mile or two one way or another doesn't mean very much until you approach your destination. And by the time you're over the airport runway, you're thinking about distance in a completely different way.

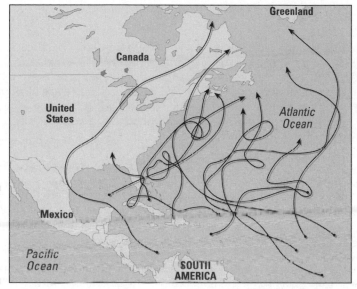

Figure 7-4: Some individual hurricane tracks.

The closer a hurricane gets to land, the more exact you want to be about the track it will take. Figuring out those next several hours, those next several miles doesn't come easily. This is not a plane skidding on the runway or a train pulling into the station. Scientists who have studied the forward motion of hurricanes say that for all the sound and fury, it moves like a fallen leaf carried along the surface of a flowing stream. And when it starts bumping into the land, the progress of a hurricane can become even more unpredictable. For one thing, the land is a new friction layer that can change the speed and direction of its winds. For another, weather systems and winds flowing from the land can complicate the picture even more.

Slowly but surely, forecasters are making progress refining their predictions of the tracks of land-falling hurricanes. On average, errors in the 24-hour track forecasts have been dropping about 1 percent per year, according to a recent study. This doesn't sound like much, but it's the kind of thing that accumulates impressive gains over the years. In 1970, the expected error in the 24-hour forecast was 140 miles, and now it's down to about 100 miles.

And the pace of progress may be picking up. "Hurricane hunter" airplanes now drop instruments through the storms that use the satellite-based Global Positioning System (GPS) to get a more accurate fix on the hurricane. And a new computer model that combines ocean and atmospheric data promises to more accurately mimic the complicated feedbacks inside a storm that affect its track and intensity.

In Harm's Way

When a hurricane is threatening a populated stretch of coastline, weather forecasters in the United States can find themselves in a real bind. High-priority public safety issues are making heavy demands on their science. Sure, computer models, better data, and smarter science are slowly improving their ability to figure out where the storm is going to come ashore. At the same time, however, Mother Nature and human nature seem to be making things harder for them every year.

A long-term pattern of relatively quiet hurricane activity in the North Atlantic seems to have come to an end in the mid-1990s. Climate has gone through a shift, and wind, rainfall, and ocean conditions that scientists think promote tropical storm formation are more frequently falling into place. Forecasters find themselves dealing with more tropical storms and more hurricanes in the region than they have seen since the 1960s.

Meanwhile, all through those quiet decades, the populations of coastal communities from Texas to New England have grown dramatically. The world over, people like to live along coastlines. So even while the number of lives lost to hurricanes in the United States has declined over the years, now the stakes are much higher than they used to be. Many more people are in harm's way. And much more expensive real estate development is at risk. Figure 7-5 shows these trends in lives lost, population growth, and property damage from land-falling hurricanes in the United States.

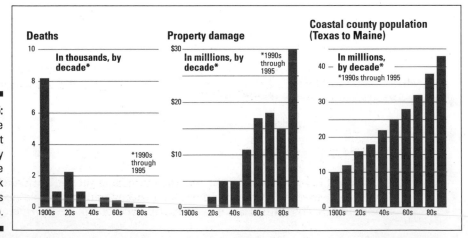

Figure 7-5: Deaths are down, but property damage and at-risk populations are way up.

Today, nearly 50 million people live along the hurricane-prone coastlines of the Gulf of Mexico and Atlantic Ocean, and the population continues to climb. And experts estimate that as many as 90 percent of these people have never experienced the real destructive power of a major hurricane. In addition to these permanent residents, thousands of vacationing tourists swell the population that is in harm's way.

Barrier islands

Of special concern to weather scientists and public safety officials are those folks occupying the low-lying strips of sand that form a protective fringe of barrier islands off the Atlantic and Gulf coasts. Many of these islands, which are separated from the mainland by lagoons, have been extensively developed with dense populations living in condominiums and resort hotels. The seaside Florida city of Miami Beach, for example, is built entirely on a barrier island. When hurricanes come ashore, residents and their properties on barrier islands can be doubly at risk. Swelled by the winds and tides, a big storm surge can overtake these low, narrow strips of sand. At the same time, escape routes from these islands are limited, so evacuations take more time.

This added congestion along the seashore means that not only are more people at risk, but they need more advance warning than they used to require. Mass evacuations take a lot more time in densely populated areas. So the 24-hour forecast is no longer enough. The public wants longer term forecasts, and of course, they want them to be accurate. But there is a trade-off about weather forecasts everywhere that is even more painfully true about forecasting hurricanes — the longer the forecast period, the bigger the errors.

When a hurricane is coming your way, here are the two levels of public announcements you are going to hear:

- ✔ **Hurricane Watch:** Hurricane conditions are *possible* in the area, usually within 36 hours. Prepare to take immediate action to protect your family and property in case a Hurricane Warning is issued.

- ✔ **Hurricane Warning:** Hurricane conditions are *expected* in the area, usually within 24 hours. Complete all storm preparations and evacuate if directed by local officials.

Playing it safe

Put yourself in the shoes of a top weather forecaster at the National Hurricane Center in Miami as an intense hurricane approaches a populated stretch of U.S. coastline. Here's the situation you face:

You have so many people at risk. They need so many hours of lead time to secure their property, to gather their belongings, and to safely evacuate their communities before the winds get too strong. You have only limited confidence in your ability to be certain about exactly when and where the big storm is coming ashore during the next several hours, and how powerful it is going to be at the time. Naturally, you want to protect everyone from the dangers ahead.

Under the circumstances, you are inclined to build a large margin for error into your calculations — to play it safe for everyone involved. So your impulse is to designate an area to receive emergency warnings and evacuation procedures that are much bigger than the storm. In the name of public safety, you are being especially cautious and covering every possibility. Who is going to fault you for being *too safe?*

The overwarning problem

After the passage of a major hurricane, two groups of people are interested in the value and accuracy of the weather forecasts and the warnings and evacuation orders that were issued. There is the group whose homes and businesses were damaged or destroyed by the big storm and who may very well have been saved from injury and possibly even death by the action that was taken. And then there is everybody else. While the smaller group of victims is preoccupied with problems of damage and repair, everybody else has time to contemplate the costs and inconveniences of the big evacuation.

Hurricane warnings come at a high price in lost economic activity as well as direct costs of damage preparation, transportation, and other public safety expenses. According to the American Meteorological Society, estimates vary between $500,000 and $1 million *for every mile of coastline* covered by hurricane warnings issued in the United States. At the same time, the state of the science of hurricane forecasting requires that warnings issued 24 hours in advance of a hurricane coming ashore covers an average of 400 miles of coastline. And yet, the typical damage path is about 100 miles. This means that three-quarters of the area of a hurricane warning is "overwarned." On average, hurricane warnings cost the United States $400 million a year. The price of "underwarning" — of so reducing the area of warning that the track of destruction falls outside of it — is increased property destruction, and almost certainly more injuries and deaths.

There may be another cost of overwarning that is harder to estimate than the economic burden it imposes. As longer periods of lead time have been required to evacuate heavily populated areas, the extent of overwarning has been on the rise. Overwarning translates into more false alarms for most people who are asked to respond to hurricane warnings and to evacuate their homes. Some weather forecasters and emergency managers worry about the public's reaction to this trend of more frequent false alarms. They wonder, Will the public lose confidence in the hurricane warning system, and so fail to respond to future warnings?

Galveston 1900

The people of Galveston, Texas, had no idea what was about to hit them. Storm warnings had been posted by the Weather Bureau for the Gulf Coast, but there was no hurricane in the forecast for Galveston for September 8, 1900. In 24 hours, at least 8,000 people were killed, and thousands more were injured and made homeless in the most deadly natural disaster in U.S. history.

The most lethal hazard of the hurricane was not the winds. Most people died from drowning in the storm surge. Galveston was built on a low-lying strip of sand known as a barrier island, and the city was completely swamped by the great tidal bulge that was swept ashore by the winds.

The true death toll will never be known. Whole families were wiped out. Thousands of bodies were strewn through the wreckage, and for days afterward they were piled up and burned.

Hurricane Force

Hurricanes are rated according to the speed of their winds. This is the hurricane's intensity, and even though storm surge and rainfall have proven to be more lethal hazards, wind speed is taken to be the measure of a storm's power to cause destruction over the countryside.

The Saffir-Simpson Hurricane Scale (see Table 7-1) is used to rank the storms according to their wind speeds, and certainly, the hurricane force of winds is nothing to sneeze at. But it is not always the best way to measure the danger of a hurricane. For example, a Category 4 hurricane is seen as more dangerous than a Category 3 hurricane, although it is not necessarily a larger storm. Wind speed or intensity apparently doesn't have much to do with size. A big storm with slower winds could endanger many more people than a smaller, more intense hurricane. Also, the pace of a storm's forward progress sometimes can be more important than the speed of the winds whipping around it. A hurricane that slows or stops in its track over an area of land can do far more damage than a more intense storm that moves through at a clip. In fact, a tropical storm does not even have to reach the intensity of hurricane-force winds to seriously threaten people, most often with flooding.

Hurricane intensity is another one of those slippery problems in weather science. As the saying goes among researchers, it is not well understood. The speeds of the winds always seem to be changing — slowing down here and now, speeding up there and then. Sometimes weather scientists can point to such things as changes in ocean surface temperatures as explanations for these shifts in wind speeds. Other times, they don't have a clue.

Predicting intensity is a major focus of hurricane research. Weather forecasters have only to remind themselves what happened with Hurricane Opal in 1995 to know how big a problem they have and how far they are from solving it. It was the kind of experience that gives hurricane forecasters nightmares.

On October 3, Opal was a Category 1 storm, with 90 mile per hour winds, drifting slowly northward through the Gulf of Mexico toward Pensacola, Florida. Forecasters went to bed that night confident that Opal would make landfall as a Category 2 hurricane the night of October 4.

Overnight, however, Opal jumped to a Category 4 hurricane with winds of 150 miles per hour and tripled its forward speed, accelerating toward the Florida coast at 20 miles an hour. Suddenly that morning, forecasters not only had a dangerous hurricane on their hands, but the evacuation time had been cut. Building toward Category 5, Opal was coming ashore at 5 p.m. that day. Fortunately, Opal faded just as suddenly to a weak Category 3 hurricane before making landfall. Opal caused several deaths and $3 billion in damage.

Table 7-1	Saffir-Simpson Hurricane Scale	
Category	*Wind Speeds*	*Likely Effects*
1	74–95 mph	Damage limited to unanchored mobile homes, shrubbery, and trees. Some coastal road flooding and minor pier damage.
2	96–110 mph	Some roofing material, door, and window damage to buildings. Damage to vegetation, mobile homes, and piers. Small crafts in unprotected anchorages break moorings.
3	111–130 mph	Structural damage to small residences and utility buildings. Mobile homes are destroyed. Coastal flooding destroys smaller structures and larger structures damaged by floating debris. Possible inland flooding.
4	131–155 mph	Extensive wall failures and some complete roof structure failure on small residences. Major beach erosion and damage to lower floors of structures near the shore. Terrain may be flooded well inland.
5	155 or higher	Complete roof failure on many residences and industrial buildings. Some complete building failures. Small utility buildings blown over or away. Major damage to lower floors of all structures near shoreline. Massive evacuation of residential areas may be required.

Hurricane Andrew

Hurricane Andrew in 1992 was the most devastating hurricane to strike the United States. The vital statistics say a lot about the two big changes in the way hurricanes are affecting the U.S.

✔ Better forecasts and warnings are saving many lives. While 53 people died, the death toll was much smaller than it would have been years ago.

✔ But many more people are living in harm's way — property damage came to about $30 billion. It was the most costly natural disaster in U.S. history.

Andrew came ashore the morning of August 24, south of Miami Beach, Florida. It took three hours to cross extreme southern Florida in the morning, completely levelling the town of Homestead. It continued west into the Gulf of Mexico, and two days later took a sharp northward turn into Louisiana.

Wind gusts were clocked at 175 miles per hour, but sustained winds measured 145 miles per hour, making Andrew a Category 4 hurricane. Its storm surge was 17 feet into Biscayne Bay. Unlike many hurricanes, Andrew caused more damage from its winds than its storm surge.

Andrew destroyed or damaged more than 200,000 homes and businesses. It left more than 160,000 people homeless. One million people in Florida and 1.7 million in Louisiana and Mississippi were evacuated from their homes.

For all its power and destruction, Andrew was a relatively small hurricane, its rainbands reaching out only 100 miles from the eye.

And forecasters say that if its track had been just a little different, things could have been far worse. Andrew's eye passed about 25 miles south of Miami Beach. If Andrew had come ashore just 20 miles farther north, according to two estimates, property damage would have been more than $60 billion.

Part III
Some Seasonable Explanations

The 5th Wave   By Rich Tennant

"Sorry I'm late, but I had to shovel about 8 inches of 'partly cloudy' out of my driveway this morning."

In this part . . .

You're going to need some good clothes. You're going to need an umbrella and some warm boots. And then you're going to want some sunscreen and a pair of cool shorts. Bring along a rain hat and a sun hat, a heater and a fan. This part takes you through the seasons of the year, and in the United States, that means you have to be prepared for all kinds of weather extremes.

The big storms of winter are about to hit. And then comes the tornadoes and all the violence of spring. Summer comes with its thunderstorms across the East and the Midwest and its heat waves that can strike just about anywhere. And in autumn, the leaves show their brilliance, and fogs often form.

Chapter 8

The Ways of Winter

Most everybody knows winter. They may be a little vague about what to expect of autumn from one year to the next, and spring can be pretty much anybody's guess. Most years, as the seasons go, winter has a personality that is hard to mistake. But the funny thing is, more than any other, winter is a season with multiple personalities. Not only is it different from year to year, but it shows a different face to every region in the country. Around the United States, this winter that we are so familiar with is so different from place to place, it is hardly recognizable as the same season.

In some parts of the country, folks are hunkering down for the winter before a lot of people elsewhere have given it much thought. Eventually, of course, the season gets most of the country in its clutches, one way or another. Not everyone is equally impressed, so different are the ways of winter. To some, it is a life-threatening experience that has literally run them out of town. To others, winter is merely an inconvenience. And there are people in southern California, basking in their desert climate, who think of winter as a television show or a place you go visit.

This chapter gets into the nitty-gritty details of the season. It describes the making of the winter storms known as *mid-latitude cyclones* and how they affect different regions of the country. It examines the season from San Diego to Seattle and from Miami to Maine. The Big Picture reason for the seasons — the tilt of the Earth — is something Chapter 2 describes. And Chapter 6 goes into the climate conditions, such as El Niño, that make one winter in North America different from the next. This chapter holds winter up to the light and the magnifying effects of a mirror that is used for shaving or putting on makeup and watches its many faces change. It's a fine season, mind you, and I wouldn't want to go without it, but you have to admit, sometimes winter is a little crazy in the United States.

Winter's "Official" First Day

No wonder winter seems a little crazy! Look at what happens when December is three weeks old: The snow has been falling for who knows how long. Where it hasn't been snowing, the rains have been coming in. It's colder than the dickens in many parts of the country, and it's beginning to feel like it's been like this forever. And along comes somebody to tell you on December 21 that the "official" first day of winter has arrived. I don't know about you, but no matter what kind of winter I'm in for, I've pretty much got the hang of it by the time December 21 rolls around. So what goes on here? Is this government work? The truth is, there's nothing official about it, and as you may suspect, it has nothing to do with weather here on Earth that day.

This is an astronomer's idea. It has to do with Earth's tilt that Chapter 2 describes. On this day, the effect of Earth's tilt is at the maximum — or more accurately, the minimum (see Figure 8-1). On December 21 or 22, the Sun's heat energy reaching the lands and waters of the Northern Hemisphere is at its lowest of the year. Often, this is referred to as the "shortest day," meaning not that it has fewer hours, of course, but that it has *fewer hours of sunlight* than any other day. Since June 21, the arc of its daily pass through the sky has been dropping farther and farther toward the southern horizon, and on this day, the *winter solstice,* it stops. The Latin word *solstice* means "Sun stands still." It has gone as far south as it's going to go.

Figure 8-1:
Earth's tilt gives the Northern Hemisphere minimum exposure to sunlight on Dec. 21.

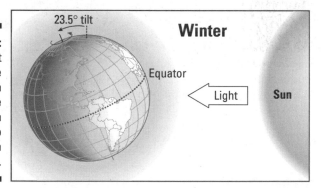

Why anyone would think of this as the *first* day of winter, I have no idea. But let the astronomers think what they will. Astronomers will tell you, by the way, that the Northern Hemisphere's winter lasts precisely 88.99 days. At the Go Figure Academy of Sciences, some of my best friends are astronomers, but, you know, I give them space. These are people with stars in their eyes. When it comes to weather, I say go with the people who have their heads in the clouds. The weather scientists know, and you and I know, how long winter has been around by December 21. We feel it in our bones.

It's a Temperature Thing

Winter is when it gets cold. As the arc of the Sun has slanted farther and farther south, the polar air masses have been migrating deeper and deeper down over the Northern Hemisphere's middle latitudes, where most people live. This migration of cold air is what makes winter what it is: a battle of warm and cold air masses. The battleground stretches clear across the country. What you feel in your bones is the fact that for quite some time there, the cold air has seemed to be winning every battle, gaining more and more territory over the warm.

If you had to pick a single day, people in most areas in the United States would say that winter begins December 1 and continues through the end of February.

In the coldest parts of the country, however, the people who really know winter better than anybody say it begins when the daily average temperature falls below freezing (see Figure 8-2). That would be the last ten days of November over most of the north, but in places like northern Minnesota and North Dakota, it would be the first ten days of November. Across a big swath of the country from eastern Massachusetts to Arizona, these temperatures aren't reached until December 15. And in much of the South and the West Coast, those kinds of temperatures are never reached — daily average temperatures never fall below freezing.

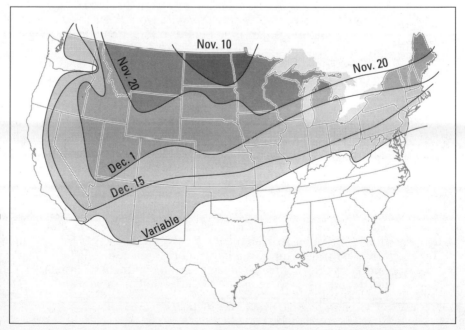

Figure 8-2: Dates when 32-degree tempertures usually arrive.

Does winter begin when the snow first flies? It does for some folks. Nothing more dramatically changes the feeling of the season or more beautifully transforms the landscape than the first layer of snow to cover the ground. Along with this big shift comes the eager hope that a good skiing season is on the way. The first snow falls as early as September in the Rockies, in October or early November in the northern and mountain states, or late November or early December in the stretch from New England to the Mississippi River.

Every winter season has its own personality, and hardly a season goes by without some surprising shifts in patterns of temperatures or storms. Some winters come early, and some years it seems to be waiting for Christmas. Many years, in many places across the country, the pattern of winter has shown its face by late November. The major storm tracks have become established across the northern tier of the United States and are populated with systems bringing snow and wind to the upper Midwest, the Great Lakes, and Northeast. Along the West Coast, the track of wet and windy Pacific storms has pushed south into California.

Coast to Coast

The fact that winter is colder and longer in the northern United States than it is in the southern part of the country is not the only important difference in the season, but it's a good place to start. After all, if it were not for this uneven distribution in the heat energy from the Sun between the Tropics and the polar regions, *Weather For Dummies* would be a pretty thin book. No other season brings out these differences as dramatically as winter.

North to south across the country, the pattern of January's average temperatures are about what you would expect. Up in International Falls, Minnesota, for example, January temperatures average 3 degrees. Down in New Orleans, Louisiana, it's likely to be 55 degrees on the same day. Out on the West Coast, the same pattern shows up: An average January day up in Seattle will see temperatures hit 44 degrees, while down in Los Angeles, it's 64 degrees. But notice something else: The huge difference in temperatures between the northern cities of Seattle and International Falls. Something else is at work.

Between the western third of the United States and the eastern two-thirds is a world of difference when it comes to weather. Think of it this way: In the western third, winter weather is controlled by the prevailing westerly winds that come riding in from the Pacific Ocean. The key is the ocean. In the eastern two-thirds, winter weather is more often controlled by clashes over the middle of the country between the cold continental air masses from the north and the warm moisture from the Gulf of Mexico and the Atlantic. The key is the land.

In most of the West, most of the region's precipitation arrives during winter (see Figure 8-3). East of the Rockies, precipitation is spread more evenly across the seasons.

Here's a thumbnail look at what winter *usually* is like — coast to coast — in different regions around the United States:

- In the *Pacific Northwest,* from western Washington, down through Oregon and far northern California, winters are rainy and mild along the coast and snowy and cold in the mountains. Farther south through California, winters steadily get warmer and less rainy.

- In the *Southwest,* the great American Desert, winter is a time of only occasional rain, long spells of fair weather, and almost always mild temperatures, although high elevations see cold temperatures and snow.

- The *Great Basin* between the Sierra Nevada and the Rockies feels cold, dry winters in the north, and warm and dry winters in the south. In the rainshadow of the Sierra Nevada, it's annual rainfall averages only 6 to 12 inches.

- In the *Rockies,* along the great mountain range that divides the country from Canada to Mexico, the western slopes wring still more moisture out of the Pacific storms flowing in from the west. Valleys are dry and cold, and high elevations get heavy snowfall and cold temperatures.

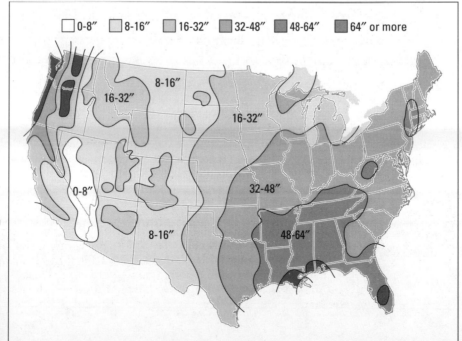

Figure 8-3: Annual amounts of precipitation throughout the United States.

- Across the *Great Plains,* the vast stretch of prairie and grassland feels the windy, frigid blasts of polar air from the north. The southern Plains get storms when this cold mixes with warm, moist air from the south. The northern Plains can get blizzards from intense low pressure that develops east of the Rockies.

- From the *Midwest* to *New England,* what most often falls from the winter storms is snow, which covers the frozen ground for several months. Here are the most storms and the longest and often the most ferocious winters of any region in the United States.

- The *Eastern Seaboard,* from Virginia to Massachusetts, may see more varieties of the winter than any region in the country. Some years, it's more often on the "warm side" of storms that ride up through the Ohio Valley. Other years, more often it's on the "cold side" of storms that come up the Atlantic Coast, carrying heavy snow and moisture from the Gulf of Mexico.

- Winter in the *Southeast,* from the Carolinas to Louisiana, is usually mild and relatively dry, as the seasons go in this warm and rainy region. But hardly a winter goes by without a few cold snaps, occasional snowfall, and even ice storms.

Storms of Winter

The Sun's role in the progress of winter is as regular and reliable as Earth's orbit, of course, but the movement of cold northern air farther and farther south over the Northern Hemisphere is not so evenly paced. It starts like a series of midnight raids and ambushes, gaining ground on a town here, a county there, and breaks out into full-scale nationwide warfare before winter is done. Farther and farther south it comes, entrenching itself behind the fronts of battle. By about the third week of January, on average, the cold air and its jet stream and the storms have migrated as far south as they are going to go. These are generally the coldest days of the season and often are the stormiest.

The storms of winter that get the whole country's attention are the sprawling giants that sweep across the United States in a matter of days before they finally trail out into the North Atlantic. They are bigger than any other storms. Every region of the country feels the effects of the biggest dead-of-winter systems. Many days of winter, on any weather map, you can see their angled fronts, like limp legs, 1,000 miles long, scissoring across the landscape. When they're talking among themselves, meteorologists call them *mid-latitude cyclones.*

Mid-latitude comes from their location, of course, which Chapter 2 describes. (If you live in the continental United States, the "lower 48," it's where you and I live.) *Cyclone* comes from the flow of the winds rotating around their centers of low pressure. This coil combines with the sharp edge along the backside of their long cold fronts to form a giant comma of cloud cover that often is the signature of mid-latitude cyclones. The rotation of a cyclone is counterclockwise in the Northern Hemisphere, clockwise in the Southern Hemisphere.

Weather scientists have three ways of looking at a mid-latitude storm. These different pictures are looking at the same weather events from different angles. What they have on their hands is a very large and complicated storm.

By coming at it from three directions, they can focus on different features that can help them get a handle on this big mess of winds and clouds, rain and snow.

- The polar front theory describes best what happens on the ground, or near the ground, when different air masses come together.

- In the upper atmosphere, scientists picture pressure waves, jet streams, and flows that promote ventilation and help these big bruisers grow.

- To put the picture together and to best illustrate the movements of winds, they build a conveyor-belt model.

Cyclones

So, what's a *cyclone?* The answer to that question depends on where you ask it.

People who live in the South Pacific and the Indian Ocean use the word to describe the storm that people in North America and Europe call a hurricane. This is the same storm, by the way, that people in the western Pacific Ocean call a typhoon.

When meteorologists use the word cyclone (as in *mid-latitude cyclone*) or cyclonic (as in *cyclonic*), here's what they mean:

A cyclone is a low-pressure weather system with a circular pattern of winds. North of the Equator, a cyclonic wind pattern is counterclockwise, and south of the Equator, it is clockwise. Take these spinning storms and slip them over their poles north and south like beanies, and what they have in common suddenly is clear. They are cyclonic because their winds are blowing in the same direction that the earth is rotating.

Their winds flow around and inward toward the center of low pressure. The winds rotate outward from a center of high pressure — an anticyclone.

A storm is born

During the months of the Northern Hemisphere's winter, the cold polar air masses and the warmer subtropical air masses to the south often tangle with each other over the middle latitudes. It is the time of greatest temperature contrasts in the upper atmosphere over this region because the polar air is plunging farther south toward the warm Equator than any other time of year.

The upper atmosphere this time of year is a shifting pattern of very large waves that mark the boundary between the air masses. On one side is the cold, dense polar air in a region of high pressure. On the other side is the warm, lighter subtropical air in a region of low pressure. At the seam between the sharply different masses of air, the jet stream and rest of the upper level westerly winds are racing along, faster and more powerfully than any other time of year.

This is where winter's storms are born, up here in the winds of the middle atmosphere, about three miles up. As the jet stream rides up over the ridges around the high pressure and plunges down into the troughs around the low pressure, it twists and changes its speed and altitude — slowing down between a ridge and trough and speeding up between a trough and ridge. Sometimes these changes make air pile up, or converge, where it slows down, and spread out, or diverge, where it speeds up. Down below, pressure builds higher under the descending air over *convergence* and surface air pressure falls in the rising air under upper-level *divergence*. When temperature and other conditions are right, this rising air generates a storm. This pattern is what weather forecasters are talking about when they say a storm has *upper air support*. Figure 8-4 is a satellite view of such a storm.

Working out the wrinkles

Around a growing low pressure system, circulating flows of warm and cold air form a big wrinkle in the boundary between the polar air mass and the subtropical air mass. A mid-latitude cyclone, a winter storm, is the atmosphere's way of working out the wrinkle.

Here are the main features of a typical winter storm:

- **Warm front:** Flowing around the low pressure, warm air pushes up into the cold polar air along a slow-moving warm front that forms a broad band of layered clouds and precipitation. North and northwest of this warm front is a broad area of rain or snow, depending on local temperatures.

✔ **Cold front:** Pushing into the warm air, the cold polar air forms a faster traveling front that is a more sharply defined and narrower band of vertical cumulus clouds, bringing intense showers and thunderstorms.

✔ **Occluded front:** As the storm matures, near the low pressure center an occluded front forms where the faster-traveling cold front overtakes the warm front, lifting the warm air off of the ground. This front can be an area of intense storminess.

✔ **Dry tongue:** A region of cold air and higher pressure and clear or clearing skies extends up the backside of the of the cold front's cloud band. The low pressure draws the tongue farther and farther north and toward the center as the storm matures. Eventually the dry tongue penetrates to the center of the low pressure and helps dissipate the storm.

Much of the precipitation from such a storm falls in the region north and west of the low pressure center which never sees the passage of the storm's warm and cold fronts. Figure 8-5 illustrates these fronts and their cloud and precipitation patterns.

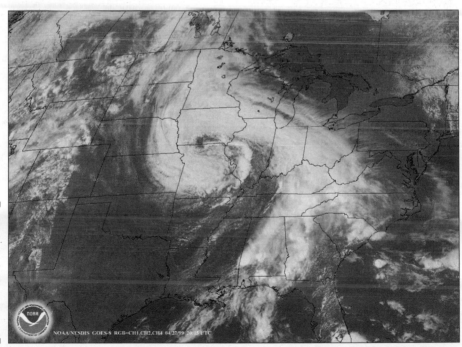

Figure 8-4:
A satellite image of a mid-latitude storm's big "comma cloud" pattern.

National Oceanic and Atmospheric Administration/Department of Commerce.

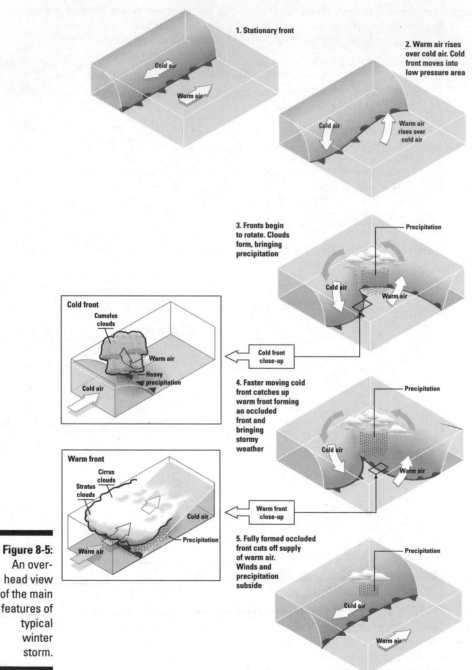

1. Stationary front

Cold air

Warm air

2. Warm air rises over cold air. Cold front moves into low pressure area

Cold air

Warm air rises over cold air

3. Fronts begin to rotate. Clouds form, bringing precipitation

Precipitation

Cold air

Warm air

Cold front

Cumulus clouds

Warm air

Heavy precipitation

Cold air

Cold front close-up

4. Faster moving cold front catches up warm front forming an occluded front and bringing stormy weather

Precipitation

Cold air

Warm air

Warm front

Cirrus clouds

Stratus clouds

Cold air

Warm air

Precipitation

Warm front close-up

5. Fully formed occluded front cuts off supply of warm air. Winds and precipitation subside

Precipitation

Cold air

Warm air

Figure 8-5:
An overhead view of the main features of typical winter storm.

What puzzled Ben Franklin

In October 1743, Benjamin Franklin had high hopes for viewing an eclipse of the moon on the night of the 21st from his home in Philadelphia. But the weather did not cooperate. A storm blew in that night — a Nor'easter, in fact — and completely obscured the sky over Philadelphia. The disappointed Franklin exchanged letters on the subject with his brother in Boston, 300 miles away. Boston's sky was clear the night of the eclipse, Franklin learned, but the city was in the grips of a Nor'easter storm the next day.

Ben Franklin established from this exchange of letters the important weather fact that storms don't simply rise up and blow out in a single location — as people thought at the time — but that they travel from one place to another across the landscape. This fact may strike you as pretty obvious in this day and age of satellites and telephones and Internet communications and what-have-you, but it wasn't obvious then.

What really puzzled Ben Franklin was this: The winds raking the city of Philadelphia that night were blowing in off the Atlantic Ocean, from the *northeast*. And yet, the storm had traveled 300 miles *northeastward* to Boston by the next day. How in the world, Franklin wondered, did the storm travel all that way *against the wind*?

It took weather scientists a long time to figure it out. These storms do not travel in the directions of the surface winds circulating around their centers. They are pushed along in the direction of the winds in the upper atmosphere.

Catching the waves

Why do some wrinkles in the polar front peter out rather than form storms? Why do some winter storms grow to be such monsters? Telling the monsters from the peter-outers is one of the most difficult jobs of weather forecasting.

In addition to the winds of the middle atmosphere, forecasters also look for short waves along the fronts that can strongly influence weather taking place on the surface. Where only a few very large long waves are standing nearly permanently aloft in the atmosphere, a dozen or more short waves can be rippling along their flanks, traveling at the speed of the winds. Short waves have the same ventilating effects as the winds on storm systems, but they don't last as long. When the waves and the winds fall into place, they generate some of the most devastating storms of winter.

In forecasting a storm, another feature forecasters look for is the location and intensity of the high pressure cell ahead of it. A strong, unyielding high acts as a barrier, forcing air to rise up over it, and adding greatly to the precipitation. Look back, for example, on all the great Nor'easters, and you will find a powerful high pressure cell ahead of the storm.

Riding the conveyor belt

Surface maps of the fronts and upper atmosphere charts of pressures, temperatures, and winds overhead are valuable, tried-and-true tools of the weather science trade. But there is something wrong with this picture. They are snapshots of two very different parts on the same creature. It's like having a picture of a head and a foot. You know there is a body there between them, but it's pretty hard to draw the connections. This problem led weather scientists to look for a new way to picture a mid-latitude cyclone, and Figure 8-6 shows what they have come up with.

The conveyor belt model tries to put the storm together as a single creature. Have they succeeded? You be the judge. It still looks pretty confusing to me. It is probably worth remembering, though, that even with all the satellites and other technologies that take pictures of it, a real winter storm still looks like a pretty complicated mess of weather no matter which way you look at it. And this picture of a storm is good at explaining several important features of its weather.

The conveyor belt model is a good picture of the storm's air flows. It shows the three main flows of air that typically make up a storm and the directions they move from the surface all the way into the upper atmosphere. Two of the flows begin at the surface, and one enters the storm from the upper atmosphere.

The three conveyor belts shown in Figure 8-6 go something like this:

✔ The *warm conveyor belt* moves up out of the wedge of warm air between the two fronts and flows up through the warm front that is rising up over the colder air to its northeast. This flow generates the wide band of layered clouds and the longer bouts of steady precipitation. It moves toward the low-pressure center near the surface, but bends to the right as it rises and eventually gets caught up in the prevailing westerlies.

✔ The *cold conveyor belt* is generated on the surface, north of the advancing warm front, in the cold air mass that the warm front is gradually overrunning. Over land, this usually is an easterly or northeasterly flow of relatively dry air, although it can pick up moisture falling out of the warm front's clouds and deposit it as precipitation northwest of the low pressure center. Near the Atlantic Ocean, this conveyor belt is a marine flow that feeds moisture into a storm and gives the dreaded Nor'easter its name. At the surface, this flow is moving toward the low center, but as it rises in the atmosphere, it, too, gets caught up in the prevailing westerlies.

✔ The *dry conveyor belt* begins in the upper atmosphere, drops down toward the low pressure center, where its cold, sinking air feeds the cold front, and then regains its altitude and also heads off again as part of the westerlies. This flow helps generate the cold, clearing skies behind the cold front, and forms the dry tongue that flows toward the low pressure and helps dissipate the storm.

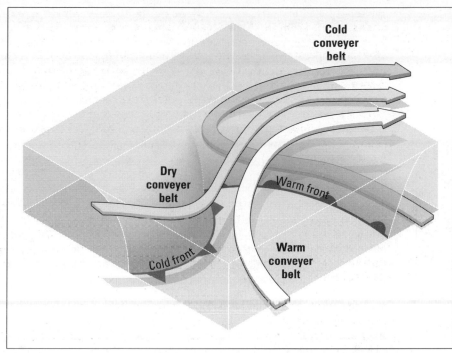

Where They Come From

Many of the winter storms that visit the United States originate in areas of the world that are much farther away than a lot of people suppose. Many of these storms are truly global in scope.

Imagine a typical winter pattern over Asia and Siberia. Over the vast expanses of snow, there builds a big high pressure system. Bitterly cold, dry air rotates clockwise around this high and circulates out into the North Pacific Ocean. There it meets and clashes with the warm, moist air flowing north along the coast of Japan above the western Pacific's Gulf Stream, called the *Kuroshiro Current*. From this contrast, a storm is born, and the prevailing westerly winds in the upper atmosphere carry it across the North Pacific.

A few days later, this storm blows into northern California, Oregon, and Washington. It dumps rain along the coast and snow over the Cascades, the Sierra Nevada, and eventually over the western slopes of the Rocky Mountains. By now, this once-powerful storm has exhausted its energy and wrung out the last of its moisture in the snowfall over the Rockies.

Just as it falls down the eastern slope of the Rockies, however, a remarkable change occurs. On the high central Plains, the old storm springs to new life as it encounters warm, moist air flowing north from the Gulf of Mexico and another frigid air mass blowing south from Canada. Now this old Asian storm takes on that raw western look of a High Plains drifter. It's big and ornery all over again, ready to kick the daylights out of the upper Midwest and New England.

A storm rebuilt like this eventually could spin out into the North Atlantic, its front closed off and its low center weakening again. Still, along the way across the Atlantic, it might pick up enough energy and moisture from the northeasterly flowing Gulf Stream to rain on England.

There are certain areas of the country where storms that cross the United States tend to develop, although as birthplaces of storms, they probably get more credit than they deserve. Rather than originating storms, often they rebuild existing storms that pass within their range. As Chapter 2 describes, these places are battlegrounds where different air masses, like opposing armies, tend to engage one another. These areas include the eastern or "leeward" side of the Rocky Mountains, the Great Basin, the Gulf of Mexico, and the Atlantic Ocean just east of the Carolinas.

Where They Go

Storms have different personalities. Some are cold, dry, and fast-moving. Others are relatively warm, moist, and slow. The amount of rain or snow they deliver depends on a variety of features. These include

- Where they come from
- The temperatures of their air masses
- The amount of moisture they contain
- Their speed of travel
- The "tracks" they take across the country

Slow-moving storms generally dump more rain and snow. Faster storms tend to generate stronger winds. As you might expect, storms with northern tracks are coldest. Southern storm tracks pick up warm, moist flows from the Gulf of Mexico and usually produce the heaviest rains and snows.

Forecasters know that where storms go, the exact track they take, can make a big difference to the type of precipitation an area might expect. The reason is this: These winter storms have a cold side, northwest of the center where the winds are from the northeast, and a warmer side, to the southeast of the center where the winds are southerly. It is not unusual for a storm to bring snow to one side and rain to the other. And notice something else: Both of the fronts are on the same side of the center. Areas on the cold side of the center may experience heavy snowfall, even though there are no fronts passing through.

Take a look at the pattern of storm tracks in Figure 8-7, and you will quickly see why New England gets such stormy winters.

Weather scientists spend a lot of time at the Mount Washington Observatory, in New Hampshire, shown in Figure 8-8. At 6,288-feet elevation, this laboratory is on the top of the coldest and windiest mountain in New England — a perfect location for studying some of the most extreme weather in the world.

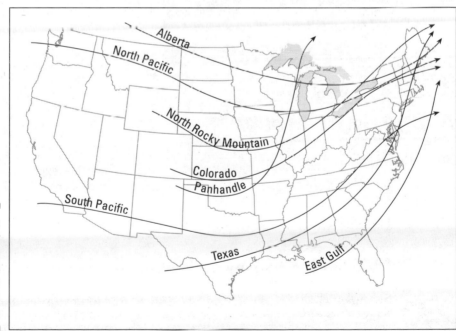

Figure 8-7:
Major winter storm tracks across the United States.

Figure 8-8:
The Mount Washington Observatory in New Hampshire.

Mount Washington Observatory, New Hampshire.

Name That Storm

Some common storms and storm tracks have earned nicknames from meteorologists around the country. These names give weather experts a shorthand way of describing a storm with certain characteristics. (Besides, they're fun.)

Here are some well-known storms that tend to visit every winter.

Alberta Clipper

An *Alberta Clipper* is a cold and windy storm that forms (or rebuilds a weakened Pacific Storm) on the eastern "leeward" side of the Canadian Rockies in the province of Alberta. An Alberta Clipper usually is a speedster that rapidly shoots across the northern tier states, over the Great Lakes, and into New England. Across the Dakotas, this storm can travel at speeds of 50 miles per hour or more. An Alberta Clipper usually does not deposit large amounts of snow, and its flakes are light and fluffy, although this storm is notorious for the strength and chill of its winds. The Alberta Clipper deserves much of the credit for Chicago's nickname, the Windy City, and its reputation for especially cold winters. Once in a while, if an Alberta Clipper travels far enough south as it moves toward the Atlantic, it can pick up enough moisture to dump heavy snow as it moves back northward near the East Coast.

Hatteras bomb

While most mid-latitude cyclones can take a week or more to grow to the size and strength of major winter storms, certain special ocean current conditions can really speed things along. These conditions develop during winter over the western Pacific's big Kuroshio Current off the east coast of Japan, and over the western Atlantic's Gulf Stream off the East Coast of the United States. Off the coast of Cape Hatteras, North Carolina, warm, moist air over the Gulf Stream can catch low pressure systems riding north along the eastern seaboard. If upper air conditions are right, and cold continental air is nearby, within 24 hours a Hatteras low can turn a young, minor disturbance into a mature, powerful monster. Many of the biggest and most damaging Nor'easters are the handiwork of these fast-growing conditions and are referred to as *bombs*.

Colorado Low

The high Plains region below the eastern slopes of the Rockies, often referred to as *leeward* slopes because they are out of the way of the prevailing westerly winds, probably generates the most winter storms for the eastern two-thirds of the United States. The region is famous for the storms it builds, but the storms are not so famous with weather forecasters, because the tracks they take northeastward are hard to predict. This information is crucial to forecasting what kind of weather a particular location, such as a major Midwestern city, can expect from them. If a *Colorado Low* passes to the north, its warm side generally brings rain. If it passes to the south, its cold side can mean heavy snowfall.

Gulf Low

Storms that form in the Gulf of Mexico and follow a track northeastward can be real winter blockbusters for the Eastern Seaboard. These systems spend most of their lives over the ocean. They soak up warm moisture and heat energy in the Gulf, cross Florida, and follow the warm Gulf Stream current up the coast. When they run into a strong high pressure cell, their moist flow can ride over the top of the colder, denser air, and forecasters can have an ornery Nor'easter on their hands. Weather forecasters can see these storms of the Gulf Low coming a lot sooner now than they used to, but still it's hard to predict how powerful they will become.

Chattanooga Choo-choo

With a little support from an upper air jet stream disturbance, a weak cold front rising up the western slopes of the Appalachian Mountains in eastern Tennessee can suddenly build into a major winter storm. The strengthening low pressure center of a *Chattanooga Choo-choo* can pull in colder air from the Great Lakes, mixing snow with rainfall. These storms generally move like a slow train to the northeast, up the Tennessee Valley and through the Ohio Valley, and they often bring long periods of heavy snow to the region.

Nor'easter

You may think that a Nor'easter comes from the Northeast. In fact, Northeasterners thought so for a long time, because that's where the wind comes from during one of these big, nasty winter storms. A *Nor'easter* is a storm that is born or rebuilt along the Atlantic Coast. Cold air clashes with the moist warmth of the Gulf Stream, the ocean current that Chapter 3 describes, and the storm rides a northern track all the way up the East Coast. The northeasterly air flow that is the cold conveyor belt of the conveyor belt model of storms is the wind that gives the Nor'easter its name. These can be big, powerful storms, like the one shown in Figure 8-9, that can bring rain, freezing rain, sleet, and heavy snowfall, as well as coastal flooding all along the East Coast. Heavy snowfall can occur as far west as the eastern Great Lakes.

Figure 8-9:
Satellite photo of a Nor'easter storm.

National Oceanic and Atmospheric Administration/Department of Commerce.

Perfect storms

If you live just about anywhere along the East Coast of the United States, you are more likely to feel the damaging effects of a big Nor'easter than a hurricane. (If you've lived there very long, you probably already know this.) A lot of people have hurricane adventures to remember, but nearly *everyone* has a story to tell about a big Nor'easter.

A hurricane is nothing to sneeze at, of course, but a Nor'easter is much more common, and its damage is more widespread. Its winds are usually not quite as fierce, but a big Nor'easter is three times bigger than a hurricane, its storm surge can be huge, and the track it travels is a weather forecaster's worst nightmare.

While an average hurricane will find a place to plow into the coast, ripping up maybe 60 miles to 90 miles of real estate, a Nor'easter can leave a trail of damage 900 miles long as it follows the coastline from Florida to Maine.

Often the slower moving Nor'easters cause the most damage. On-shore winds build a dome of ocean water that surges ashore ahead of the tempest. The longer its winds blow, the bigger this storm surge and the deeper its floodwaters are likely to be.

Nor'easters can dish out winter's worst. Their storm surges and torrential rains can cause widespread flooding. Their winds can gust to hurricane force. Their snows can bury the countryside, and devastating damage from the freezing rain of ice storms is not uncommon. Through the Southeast, they can spawn tornadoes, the storms that Chapter 9 describes.

Pineapple Express

WEATHER JARGON

Once in a while, the polar jet stream that carries winter storms across the northern Pacific to the West Coast of the United States dips down toward the tropical ocean just as a particularly stormy episode is underway out there. These tropical thunderstorms pump enormous amounts of water vapor into the atmosphere. The westerly wind drags the warm moisture across the ocean, passing near Hawaii along the way, and delivers especially heavy rains to the Pacific Coast. This is called a *Pineapple Express,* and its torrential rains are dreaded in California. The big rivers in Northern California are threatened with flooding during extreme rainfall and Southern California's desert soils are threatened with flash-floods and mudslides.

Siberian Express

When low temperature records start falling in the eastern half of the United States, look for signs of the *Siberian Express,* which is not so much a storm as it is a blast of arctic air. This frigid air mass usually forms in a large high pressure area near the North Pole or even over Arctic Siberia. Something — usually it's a jet stream disturbance over this upper air ridge — causes a portion of this dense, dry air to break away, like an iceberg from a glacier, and to slide south. As it passes over the Canadian prairies, the eastern flanks of the Rocky Mountains generally steer this shallow, heavy flow into the eastern two-thirds of the United States. As it sinks deeper and deeper south, it spreads a fast-moving front of precipitation, a sudden drop in temperatures, and bone-chilling winds. Passage of a Siberian Express can be a dangerous and damaging weather event as it moves deeper south across the United States. Its wind chills can endanger people who are unaccustomed to the hazards of such extreme temperatures, and the freezing air can cause heavy losses to such all-year crops as citrus groves.

Panhandle Hook

This storm builds east of the southern flanks of Rockies in the area of southeastern Colorado. It pokes along to the southeast, crossing the Oklahoma panhandle, picking up moisture from the Gulf of Mexico along the way, and then takes a sharp hook-turn northeastward and picks up speed. A *Panhandle Hook* plows up through the upper Midwest, out over the Great Lakes, and into Canada. This storm is notorious for delivering heavy snowfalls to northern Illinois and Wisconsin.

Texas Panhandler

Texas Panhandler is another storm that builds from the combination of cold air blowing down out of the eastern slopes of the Rockies and warm marine moisture flowing up from the Gulf of Mexico. Unlike a hook, this panhandler follows a more common track across the Central Plains and the Midwest. Especially when this storm is moving through a region of subfreezing temperatures, the warm Gulf air flowing up the warm front slope of a Texas Panhandler can bring heavy snows to a vast region of the nation's midsection.

Blue Norther

A *Blue Norther* or a *Texas Norther* is a name sometimes given to an intense storm that tracks eastward across the Great Plains. Northerly winds behind the storm can send frigid air plunging south through Texas. Temperatures across the region can quickly drop 10 degrees or more. Often such a wind is accompanied by heavy snow that can bring blizzard conditions to the southern Plains.

When the Flakes Fly

Everybody loves a snowflake. Up really close, the elaborate and lacey beauty of its construction is a marvel, to be sure. There are not very many delicate things about the ham-handed ways of winter, but Nature seems really to have outdone herself with the design of flakes of snow.

About 100 years ago, a Vermont man, Wilson A. Bentley, devoted 50 years to studying snowflakes, microscopically photographing thousands of these little beauties. The Snowflake Man, as he came to be called, claimed never to have seen two that were exactly alike. Figure 8-10 is an example of his work. Whether or not there are any two snowflakes alike, as Figure 8-11 illustrates, there are definitely distinct types of flakes.

One on one, snowflakes are fine, of course, but put billions and billions of these delicate little beauties together, and they take on a different look. The first snowfall of the season seems always to be a sensation. It takes a while for the eyes to become accustomed to its whiteness, to the way its insulating blanket so completely makes over the landscape, and for the ears to adjust to the quiet it suddenly gives the day. Rain is a brass band in comparison.

Figure 8-10:
A micropho-
tograph of a
snowflake
by Wilson A.
Bentley.

National Center for Atmospheric Research/University Corporataion for Atmospheric Research/National Science Foundation.

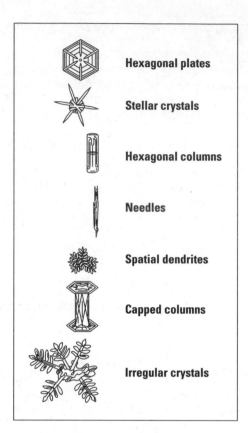

Hexagonal plates

Stellar crystals

Hexagonal columns

Needles

Spatial dendrites

Capped columns

Irregular crystals

Figure 8-11:
Common
types of
snowflakes.

On the other hand, if you live in one of the great snowbelts of the United States, it doesn't take long for the charms of snowflakes to lose their hold on you. Have you ever noticed how quickly snow picks up the dirt and grime on the side of the road? One reason is the sand and other materials that highway crews spread on the road to improve traction. Another is the soot from vehicle exhaust.

Whew! This is not something you want to take home with you. And notice this about these precious little jewels: When they gang up on you, when they fly in your cold face during a storm, most of them are ragged, broken pieces of flake. For much of the country, snow is a heavy burden.

Even in the regions of the United States that enjoy warmer winters, most precipitation begins as snow, in temperatures in the clouds that are below freezing. The difference between rainfall and snowfall depends on the temperatures of the layers of air the snow falls through on its way to the surface. More often than not, a raindrop is a melted snowflake.

Measuring snow

Measuring rainfall is usually a pretty simple matter of reading the depth on a gauge that is attached to a long, narrow cylinder with a funnel at the top. Measuring snow, on the other hand, is complicated by the flaky nature of the stuff.

In the slightest wind, snowfall is much less evenly distributed than rain. Even a halfway serious winter storm will blow it all over the place. Snowfall depth is generally measured on the top of a board that has been wiped clean since the last time of observation. To overcome the problem of the winds, weather forecasters measure the depth at several locations and then do the math to come up with an average.

This new depth figure is then added to the depth of the snow underneath the boards to come up with a total snow accumulation.

But there's another complication. In most places, it's important to know how much liquid water this amount or that amount of snow represents. As a general rule, 10 inches of snow melts down to 1 inch of water. But those figures wander all over the place. Really dry, fluffy flakes may take 30 inches to yield an inch of water, while only 3 inches of very wet snow yields an inch of liquid.

Often, there is a fairly distinct boundary between an area of rainfall and snowfall from a large storm. In large urban areas, which depend on fleets of snow removal equipment to keep commuters commuting, one of the more important and difficult jobs of weather forecasters is to accurately predict not only precipitation, but when and where snow or ice will occur.

It may be a matter of temperature, but the difference between rain and snow is not just a matter of degrees. If rain is expected, most often the most important question is, "*When* is it going to rain?" If snow is expected, on the other hand, the most important question may be, "*How much* is it going to snow?" Highway crews, on the other hand, are very much interested in when snow is expected to fall. While they fight a 3-inch snowstorm much the way they will fight an 8-inch storm, a difference in starting time between 7 a.m. and 10 a.m. is critical.

Blizzards

Along with ice storms, the snowstorms known as *blizzards* are the worst that winter dishes out. Their winds are strong, and their snow is blowing to the point where visibility is very poor. And they're colder than the dickens. The word blizzard gets kicked around a lot on nasty winter days, and often it is used by forecasters to describe a storm that brings heavy snow and extreme blowing and drifting of snow.

The National Weather Service has an official definition of blizzard as a storm with a wind of 35 miles per hour or higher, low temperatures and enough snow in the air to reduce visibility to less than a quarter of a mile.

Blizzard warnings hit the airwaves when three things are happening at once. Wind speeds are hitting at least 35 miles per hour. Snow may not be falling heavily, but a lot of it is flying in the air — usually the small, light flakes that accompany particularly cold storms and are easily picked up and blown by the winds. Temperatures, meanwhile, are 20 degrees or lower, and they're expected to stay that cold for some time.

Snowstorms can get worse. *Severe blizzard warnings* are issued when winds of 45 miles per hour are expected, heavy snow is falling or blowing, and temperatures are below 10 degrees.

Cold polar air behind the passage of a winter storm across the Great Plains can carry winds so strong that it whips the freshly fallen, fine powdery snow off the ground. This is often called a *ground blizzard,* because snow is not actually falling at the time, although the strong winds, the extreme cold, and the low visibility from blowing and drifting snow are just as real and dangerous.

Some blizzards are so bad they go down in history.

"Great White Hurricane"

One of the worst winter storms ever to hit the eastern United States was the blizzard of 1888, a storm that in the middle of March devastated a quarter of the population. Seamen who lived to tell about it called this storm the "Great White Hurricane."

Some villages and towns from Maryland to Maine were buried under 40 inches to 50 inches of snow and drifts of between 30 and 40 feet.

The cities of Washington, D.C., Philadelphia, New York City, and Boston were cut off from the outside world and brought to a standstill. Telegraph, telephone, and electrical wires broke under the weight of snow and ice and the force of the winds.

Temperatures ranged from 0 to 20 degrees over the region, and winds howled for two days at nearly the force of hurricanes.

More than 400 people died, many lost in the giant drifts in city streets. Passenger trains were buried and marooned. More than 200 ships were sunk, wrecked, or abandoned, and 100 seaman died.

In February 1899, another great blizzard hit the region and brought one of the worst cold waves in United States history. The low of 15 degrees below zero in Washington, D.C., still stands as the all-time record for the nation's capital. The mercury hit 12 below zero in Georgia, and for only the second time in recorded history, ice floated down the Mississippi River to the Gulf of Mexico.

Ice Storms

Nothing in winter does the damage of an ice storm. A blizzard may pose more perils to the safety of people while they are out in it, because winds are so strong and visibility is so bad, but it is usually the kind of weather that you can shelter yourself against. A long storm of freezing rain crushes everything under the weight of the glaze of ice it accumulates. Electrical power systems that depend on overhead lines are no match for a serious ice storm. Power lines swell with ice and droop under the added weight, and before long, their poles snap.

After a blizzard passes, you can usually dig out and go about your business. The effects of an ice storm linger for days. Things are broken, the power is off, and transportation is nearly impossible. Driving a car through snow takes patience and caution. Driving on a street glazed with ice from freezing rain is nearly impossible. It's like trying to make your way across an ice rink.

The rain from an ice storm usually is the handiwork of a warm front, but it's the layer of very cold air underneath it that deserves the blame for the ice. This frigid arctic layer is more dense — heavier — than the air around it and tends to stubbornly hang on in low-lying valleys. This is the stuff that causes the raindrops to freeze into glaze as they hit the ground.

"The Blizzard of '96"

Most everybody calls what happened on the East Coast in January 1996, the "Blizzard of '96," and certainly it's worthy of the name, although in the technical meaning of the word, it may not have had the sustained wind speeds to qualify as a blizzard.

Whatever it was, it paralyzed all major cities along the East Coast from Richmond, Virginia, to Boston. It forced the closure of airports, businesses, schools, and government offices — including the federal government, which closed for four days.

Severe thunderstorms and record snowfall from this storm also hit the Deep South.

The storm packed the entire region with snow, and then a week later, a warm rainstorm washed through much of the East with 2 inches to 6 inches, bringing rapid snowmelt that overflowed rivers from the eastern Ohio Valley to New England and south to Washington, D.C. Some 154 people died in the storm, 33 were killed in the flooding, and damage from the two events totaled $3 billion.

Life and Limb

Every season has its own set of health and safety hazards, and winter in most areas of the country certainly has its share.

Naturally, regions of the world with the hardest winters see these hazards most often, but experts have noticed something else: It's where severe winter weather is most unusual that it poses the biggest threat. People who live in places that frequently experience severe cold and heavy snows know how to handle these conditions and how to avoid threats to their safety. Basic precautions against the dangers of severe cold are part of the daily routine of life during winter in the Great Plains, the Midwest, and Northeast. In the southern United States, on the other hand, where houses aren't built for cold weather and folks are less accustomed to its hazards, the occasional ice storm is a bigger safety threat.

Severe cold can overwhelm the body's ability to produce heat. When the body feels this problem coming, the heart begins pumping faster in order to try to keep up the heat. It's working extra hard to fight off this cold, and this is the time to be especially careful about asking it to do a lot of extra physical work.

"Ice Storm of the Century"

It was Canada's worst natural disaster. Five days of rain fell in early January 1998 as warm, moist air from the Gulf of Mexico rode over a layer of Arctic air and the rain falling through the frigid air froze instantly on every exposed surface.

In the northeastern United States, federal disasters were declared in Maine, New Hampshire, Vermont, and upstate New York, but populous southeastern Canada took the worst.

From eastern Ontario to New Brunswick and Nova Scotia, the weight of the ice toppled steel electrical power transmission line towers and crushed trees. A total of eight million people were without electricity at the height of the disaster, and many were without power for a month. One climatologist described the damage to the electrical power grid this way: "What it took human beings a half-century to construct took nature a matter of hours to knock down."

The storm killed 35 people, and property damage was in the billions of dollars. In Quebec and Ontario, where 90 percent of the world's maple syrup is produced, millions of maple trees were destroyed. By one estimate, it will be 30 to 40 years before maple syrup production returns to normal.

Avoid overexertion and exposure. Exertions from attempting to push your car, shoveling heavy drifts, and performing other difficult chores during strong winds, blinding snow, and bitter cold of a blizzard may cause a heart attack — even for people in apparently good physical condition.

Avoid alcoholic beverages. Alcohol causes the body to lose its heat more rapidly — even though one may feel warmer after drinking alcoholic beverages. Never give the person alcohol, sedatives, tranquilizers, or pain relievers. They only slow down body processes even more.

Cold weather exposure

When your body loses the battle against the cold, often it's somebody else who notices it. When the cold has got you bad, you are not always the best judge of the seriousness of the problem. You still think you're okay, you just need to rest a minute longer. It's somebody else who realizes that you are showing some of these signs of *cold weather exposure:*

- You can't stop shivering.
- You're fumbling your hands.
- Your speech is slow and slurred, and maybe even incoherent.
- You're stumbling and lurching as you walk.
- You're drowsy and exhausted.
- Maybe you've rested and can't even get up.

A person acting like this needs to get into dry clothes and a warm bed. They need a warm "hot water bottle" or a heating pad or warm towels on their chest or shoulders or stomach. They need some warm, nonalcoholic drinks. They should have their feet up, so warm blood flows to their head. They need to be kept quiet. They shouldn't be massaged or rubbed. In extreme cases, they need a doctor.

Frostbite

Sometimes, when you *stop feeling cold* can be the most dangerous sign. *Frostbite* — the freezing of your body's tissue — causes a loss of feeling in your fingers or toes, the tip of your nose, or you ear lobes, and they appear white or pale.

A person with frostbite needs medical attention immediately. They should not rub the affected tissue with snow or ice. The frostbitten tissue needs to be warmed, and the person should be treated as a victim of exposure.

The best thing is to avoid frostbite. Dress warmly in loose layers of clothes covered with a water-repellant shell. Stay dry. Change those wet socks. Not only do wet clothes lose their power to insulate against the cold, but the water in them conducts the body's heat away. Wear a hat. And as most everybody in cold country knows, mittens are a lot warmer than gloves.

Caught in a car

You're caught in your car in a blizzard. Do you stay in the vehicle? Or do you give in to that panicky urge and try to make a run for it? Experts in these matters say there are no two ways about it: The only safe thing to do is *stay in your vehicle.*

Nothing is easier or more dangerous than getting lost in the whiteout conditions of blowing and drifting snow during a blizzard.

 Your car is shelter, and you are more likely to be found in it than out of it.

Here are some tips:

- Keep the vehicle's dome light on to make it more visible for rescuers.

- Watch out for *carbon monoxide,* the odorless exhaust gas that is poisonous. Run the car's engine and heater only as often as you need to. Make sure that a downwind window is open for ventilation and don't let snow block the exhaust pipe.

- Move around in the car, changing positions once in a while and doing a little exercise, like clapping your hands and moving your arms and legs.

- If you're alone, try to stay awake as long as possible. If you are not alone, sleep in shifts, keeping one person watching for help.

Chapter 9

Twists and Turns of Spring

Ah, spring! How come nobody says, *"Ah, winter"*? Among the seasons, year in and year out, spring gets the best publicity. Winter could use some help from spring's public relations consultant. Everybody knows what to call a cold and stormy and miserable winter season. They call it a *bad winter* or a *hard winter*. Spring, on the other hand, is a season that can do no wrong. What do people call a cold and stormy and miserable spring? They call it a *long winter!*

The poets have done this for spring. Those romantic homebodies have this special thing for this season, and I think I know why. Sure, everybody likes to feel that warming Sun and to see the lengthening hours of daylight. But it's not just the weather. Spring gets credit for everything — every flower bud and sprout, and the song from every robin and bluebird that returns from the south. And when those romantic poets finally pull back their curtains and get a look at what the birds and the bees are up to this time of year, well, it sets their hearts to singing.

But there's another side to this season of transition between winter and summer. Spring has a mean streak. It is not stable, and among forecasters, its weather has a reputation for being painfully unpredictable. Every season has its own perils in the United States, but there are parts of the country where spring can be a really dangerous brute. Spring is about the business of reclaiming the ground that was lost to winter's cold. It is wielding the power of an advancing Sun, and taking no prisoners along the way.

This chapter is about this *other side* — the thunderstorm side — the mean side of spring. Across the southeastern United States, the Great Plains, and the Midwest, dangerous thunderstorms can fill the spring sky. You can hear their great roars and see flashes of lightning burning hotter than the surface of a star. Spring is the peak season for tornadoes, the most violent and terrifying winds on the planet.

When Has Spring Sprung?

If you were in the business of running the Earth, spring probably would not be the first season you would want to hire. For all their faults, winter and summer are much more reliable. My advice would be to put them in charge of the place. You can't do without it, you know, but spring has this difficult, fitful temperament. Some days it does great work, and you're smelling the daffodils. The next day, it doesn't show up, and you're shoveling snow again.

And then there are its thunderstorms. Tempestuous is a word that comes to mind. Of course, everybody knows that eventually spring gets its work done. Because you're in charge of the place, you can let somebody else — some poor weather forecaster — worry about whether spring is going to show up from one day to the next, and how it's going to behave.

Astronomers see the Big Picture, and from the distance of space, everything looks pretty much like clockwork. They don't have to deal with the mess that the atmosphere creates — you know, the weather — so they have a clear and simple idea about spring.

To astronomers, the season is marked by a certain spot in Earth's orbit around the Sun. They call it the *vernal equinox,* the instant when the planet reaches the halfway mark in the annual circular path that leads from winter to summer. Chapter 2 shows how the seasons are caused by the fact that Earth is tilted in relation to the Sun. The vernal equinox almost always happens on March 21 — the day when the effect of Earth's slant is the same, *equal,* in both the southern half and the northern half. The Sun is directly over the Equator, and everywhere that day the length of day and night are equal. That's what the Latin word *equinox* refers to — this equal time. *Vernal* refers to spring. The same thing happens in September at the autumnal equinox. (So, if autumn is another name for fall, how come vern isn't another name for spring? Go figure.)

On March 21, there are 12 hours of sunlight and 12 hours of darkness. Astronomers call this the first day of spring.

Well, it's a big country, and whether or not March 21 sounds like the first day of spring to you pretty much depends on where you live. Spring comes earlier in the southern parts of the United States and later in the north. The general idea that spring is the months of March, April, and May is a commonly accepted compromise.

Coast to Coast

In the Pacific Ocean north of the Equator, the large high pressure system that dominates much of the weather in the western United States has begun to rebuild in strength in early spring, and the jet stream that guides storms into the region has begun retreating back toward the pole. Some years in southern California and the Southwestern desert, it feels like spring begins shortly after New Year's Day, although people in the Pacific Northwest have months more of rainfall to look forward to. Along much of the frostfree West Coast, days of spring usually are mild and often clear of fog. (Hey, they don't call it *pacific* for nothing.)

In the heights of the western mountains, spring is a radical mix of bursting sunlight and snow showers, which can linger as late as June. Still, the warming trend is allowing creeks and rivers to begin swelling with runoff of melting snow.

Thunderstorms, spring's storms of choice, are not uncommon in the West this time of year, especially along the western slopes of the mountains, but they are nothing like the violent giants that are taking shape east of the Rockies. From the Great Plains, across the prairies, and over the big river valleys of the Midwest, the long season of the thunderstorm has begun. Figure 9-1 shows you the "average thunderstorm days" around the country every year. Before spring is over, Tornado Alley, from northern Texas to Iowa, will begin to live up to its reputation.

In the Atlantic Ocean, the Bermuda High is beginning to gain strength and is shifting westward. Air circulating around it carries warm moisture from the Gulf of Mexico into the interior of the United States. The warm, jumpy moisture will clash with the cold, dry dense air masses left over from the continental winter. Cold and dry, warm and moist — air masses with these traits do not get along. The bigger their differences, the more severe the thunderstorms they make. For months now, the sky over an expanding region in the midsection of the country will have some nasty fights on its hands.

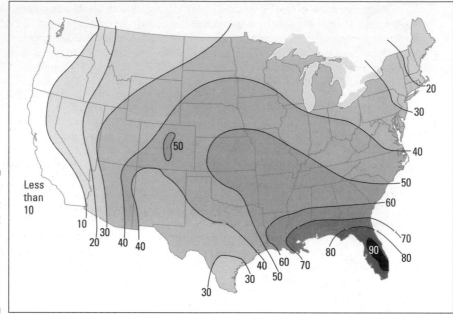

Figure 9-1:
The number
of days that
thunder-
storms are
reported, on
average,
each year.

Thunderstorms

The National Weather Service estimates that every year there are about
100,000 thunderstorms in the United States, and about 10,000 of them are
severe. Around the world, more than 40,000 thunderstorms form every day.
Nearly 1,800 thunderstorms are occurring *at any moment* around the world.
The people at the National Weather Service have done the math: That's 16
million a year!

Looking at the planet as a whole, thunderstorms are one of the main ways
that Earth pumps water vapor from the warm ocean into the atmosphere and
so begins redistributing the Sun's uneven heat energy. Chapter 2 describes
this heat-distribution problem.

Thunderstorms are a little hard to believe. For one thing, a lot of them seem
to come out of nowhere. They take shape so fast that they can be on top of
you before you realize it. And while they seem enormous when you're under
one, as storms go they are too small to be predicted by the big computer
weather models. Think about it. If you hadn't seen one, would you really
believe that a single storm could set the sky on fire with humungous flashes
of electricity, roar like a giant beast, smash the ground with big chunks of ice
on a hot summer day, and thrash the land with the most powerful winds on
the planet?

Whatever they do for the planet by helping to sort the Sun's heat more evenly, these storms are a menace to society. (And they ought to be locked up!) Just look at the messes they make in the United States alone:

- Their lightning kills dozens of people and causes thousands of fires every year.
- Their hailstones cause $1 billion a year in property damage and crop losses.
- Their flash floods are the biggest killers of all.
- They have straight-line winds that can hit 100 miles per hour or more.
- They have downburst winds that can cause airplane crashes.
- And then there are their tornadoes, 1,000 or so a year.

On the other hand, the rainfall from these storms is not something you would want to go without. It is the main source of moisture for the crops that grow across much of the United States. And their lightning flashes fix nitrogen compounds in the soil, producing a form of fertilizer.

Like all storms, a thunderstorm needs rising air. Chapter 5 describes the various ways air gets a lift off the surface. And the temperature of the upper air needs to be arranged so that the air continues rising — conditions that weather scientists call unstable.

The key to the size and power of a thunderstorm, weather scientists say, is the strength of the up-and-down motions of its air flows, the updrafts of warm, moist air and the downdrafts of cool precipitation. Weak updraft storms are more common and less threatening than strong updraft storms, as you probably figured out, and *supercells* — well, their updrafts and downdrafts are super. You can usually see the difference. A weak updraft storm makes a cloud top that is softer looking and less shapely. A strong updraft makes well-formed cauliflower edges and a big flat crown that looks like a blacksmith's anvil.

Thunderstorms organize themselves into cells. (I *told* you they should be locked up!) A cell is a single pair of air motions — an updraft coupled to a downdraft. Figure 9-2 gives you an inside look at a typical thunderstorm.

Also, midlevel winds that tilt the top of the storm allow a cell to continue to grow rather than collapse onto itself from the dampening effect of its own downdraft. The slant shape gives the cool downdraft carrying precipitation and the warm updraft carrying moist "fuel" their own spaces, separate from one another.

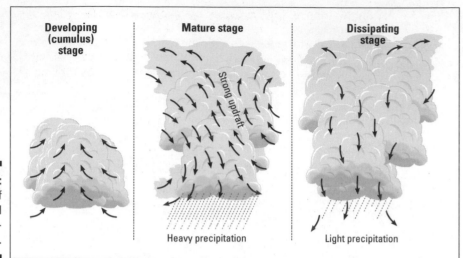

Figure 9-2:
Life cycle of a single-cell thunderstorm.

The winds can make a big difference in how a thunderstorm develops and what kind of threat it can become. When the air is fairly calm when thunderstorms come along, they can sit over the same place for hours, dumping flood-threatening gutter-gushing rain. When winds are strong, on the other hand, things can go two very different directions. The winds may blow the storms apart, or they may become powerful giants. Predicting which is going to happen is not so easy.

How . . . exactly . . . "severe"?

You hear it all the time in weather watches and warnings during spring and summer: Severe thunderstorms are on the way. The meaning is clear enough: These things are dangerous, and it's time to keep your wits about you. But what exactly is so severe about them?

What the National Weather Service means by *severe* — as in BULLETIN: SEVERE THUNDERSTORM WATCH — is a storm that produces *one or more* of the following safety hazards:

- ✔ Hailstones that are ¾-inch or larger.
- ✔ Wind speeds that are 58 miles per hour or faster.
- ✔ Tornadoes of any size.

A *watch* is an alert that conditions in a large area are ripe for severe thunderstorms to form. A *warning* comes when severe thunderstorms are in your area now. Don't wait. Go to a safe place. When you hear a warning, don't get caught watching.

For a thunderstorm to become really severe, and dangerous, several things fall into place:

✔ A big fuel supply of warm, moist air keeps pouring up into the sky.

✔ A supply of dry, thirsty air that sucks up the moisture through evaporation. Between the dry and moist air can form a dryline of plunging air.

✔ Upper air temperatures remain unstable, so that rising air is always warmer than the air around it and so it continues to rise.

✔ Winds are blowing in different directions, causing the thunderstorm cells to tilt over, giving their updrafts and downdrafts their own spaces to grow.

✔ If a layer of warmer air above the ground is keeping a lid on things a while, once the rising air breaks through this *convective cap,* everything explodes upward.

Shapes and sizes

Thunderstorms come in different shapes and sizes. The smaller ones are less severe, as a rule, although even a modest-sized storm can cause heavy rains or hail and dangerous lightning. And their downbursts can be a threat to airplane pilots.

And the supercells, well, they're in a class by themselves.

Single cell

A *single cell thunderstorm* usually lasts between 40 to 60 minutes. They can produce downbursts, hail, heavy rainfall, and once in a while even a weak tornado, but most single cell storms do not produce severe weather. In fact, a completely isolated single-cell storm is really kind of a rare creature. The conditions that make a single storm usually can just as easily make more than a single storm. And thunderstorms that seem to be separate single cells can be linked by their updrafts and downdrafts that feed on one another.

A typical single-cell storm is relatively brief because winds are usually light. The storm stands up straight under these conditions and before long the cool downdraft collapses directly onto the warm updraft. As Figure 9-3 illustrates, the big difference between such a minor thunderstorm and a severe thunderstorm is in the tilt. If the winds can push the storm over at an angle, the downdraft will fall ahead or behind the updraft, and the cell will continue to grow.

Air mass thunderstorms is a name that is sometimes given to common single-cell or multicell thunderstorms that rise quickly in the heat of the afternoon and dissipate in about an hour. They are not related to a front of air or other weather system. These types of thunderstorms are especially common in the humid summer days in the southeastern U.S., especially Florida.

Jim Reed.

Figure 9-3:
A single-cell
thunder-
storm.

Clusters

The most common of the storm types, *multicell clusters* are groups of cells in different stages of development, moving as a single unit, and capable of moderate hail, flash floods, and weak tornadoes. Forecasters know immediately that a cluster is much more likely to produce severe weather than a single-cell storm.

In a multicell storm, as shown from the inside in Figure 9-4 and from the outside in Figure 9-5, the clouds along the leading edge of the storm will show signs of decay, their wind-driven anvil tops pulled out ahead of the cluster. The stronger updraft clouds will be near the center. A flanking line of developing clouds brings up the rear, facing the wind, which is usually from the west or southwest.

Typically, rain or hail falls from the highest towers. Some clusters are well-organized systems that can produce severe weather, while others are not so likely to do so. The threat of flash flooding is much higher from multicell clusters than single-cell storms. Clusters can also produce hail the size of golf balls and winds up to 80 miles per hour.

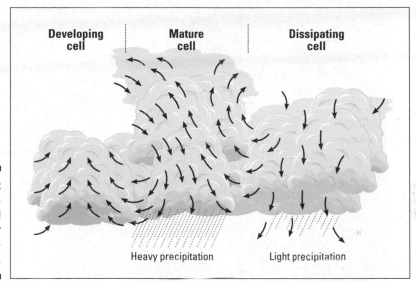

Figure 9-4:
Inside a
multicell
cluster
thunder-
storm.

Developing cell   Mature cell   Dissipating cell

Heavy precipitation   Light precipitation

Figure 9-5:
A multicell
cluster
thunder-
storm.

Jim Reed.

Complexes

Thunderstorm cells also can organize themselves into large circular cluster
systems called *mesoscale convective complexes.* (Mesoscale is a five-dollar
word for something that is usually a couple hundred miles across.) A *squall
line* is a mesoscale complex, but they can also take shape in large circular
systems, as the satellite image of Figure 9-6 illustrates.

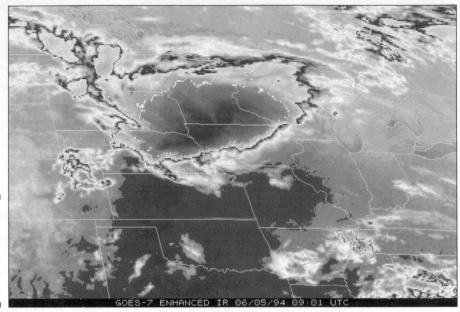

Figure 9-6:
A satellite
image of a
mesoscale
convective
complex.

Jim Reed.

As their downdrafts and updrafts feed off each other, dying off in one spot and building up in another, these systems also develop other features that help keep them going for hours. The cold, dense downdrafts force warm moist air upward, and as more and more of them begin working together, such a system can develop its own low pressure at the surface, causing air to rush inward and adding to the rising motion.

These complexes are great pulsating circles of thunder and lightning, rain and wind. Often forming at night, such a complex can spread over hundreds of square miles, covering a whole state at a time. They last for 12 hours, on average, but they can last 24 hours. Tornadoes can form, but the biggest threat from these complexes is flooding. They can bring hours of heavy rain to an area. According to one estimate, these complexes account for 80 percent of the rainfall supply during the growing season in the Midwestern cornbelt.

These systems can begin taking shape in late afternoon or early evening in the Great Plains or the Missouri River Valley and tend to reach their peak during the late night and early morning hours. As they build in strength they tend to migrate eastward, sometimes reaching the Great Lakes or the East Coast.

Squall lines

Thunderstorm cells can organize themselves along a narrow line that runs for hundreds of miles in front of an advancing cold front. *Squall lines* can form up against a cold front, where the dense cold air is plowing under warm moisture, making it rise. Also, somehow, they can develop as far as 100 miles out ahead of the front in the warm, moist air mass of a middle latitude storm, which Chapter 8 describes.

Early in the day, squall line thunderstorms typically form near the front. Later in the day, the building downdrafts from the front pushes the line farther east. Helped along by the rising air motion during the heat of the day, squall lines ahead of cold fronts usually reach their peak strength in late afternoon and evening. Usually they weaken overnight as surface temperatures cool.

Their cells have a somewhat different look to them. While many advancing thunderstorms dump heavy downpours of rain or hail in front of them, as Figure 9-7 illustrates, squall line cells lead with their winds.

Figure 9-7:
Inside a squall line thunderstorm.

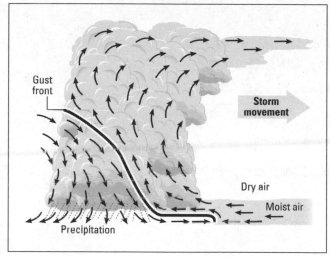

A dark bank of clouds coming out of the western sky, a squall line arrives with a bang (see Figure 9-8). Its long straight-line gust front of in-rushing warm, moist air can reach 80 miles per hour, hurricane force, and its powerful, cold downdrafts can come with sudden bursts of rain or hail. Weak tornadoes can form along squall lines, but their greatest threat is flash flooding. Driven by an advancing cold front, a squall line usually moves fairly quickly through an area, although layers of heavy clouds in its trail can bring hours of heavy rain. And when their frontal system stalls, squall line cells can drift back and forth, bringing torrential rains.

An approaching squall line can play tricks on you. Looking at the dark clouds leading the way, you might think the storm will blow over once they pass. But often behind the dark roll is heavy rainfall coming from the lighter clouds in the milky sky that you can't see so well.

Airports and airlines keep their eyes peeled for these great walls of violent winds. Pilots can find themselves diverted hundreds of miles off of their normal flight paths in order to avoid a big bank of 50,000-foot-high squall line clouds.

Figure 9-8:
Satellite
photo of a
squall line of
thunder-
storms.

Supercell

A giant tornado machine known as a *supercell* is a thunderstorm with a deep rotating updraft. Its cloud, a cumulonimbus described in Chapter 5, reaches up ten miles or more through the whole layer of atmosphere where weather takes place. While clusters and squall lines are organized into separate cells that move together, a supercell operates more or less like an economy-sized package of a single-cell storm — run amok!

There is also a key difference in the *rear flanking line,* the part of the storm that looks like the tail. In a cluster, this side facing the wind is generating updrafts that will make new cells that compete with the old cells. This single-cell giant, on the other hand, doesn't put up with competition. The updrafts of the flanking line are pulled directly in to feed the updraft of the supercell.

A supercell gets its updraft rotation by tilting the winds. This rotation speeds up the updraft, causing the cell to grow bigger faster. When the air comes rushing in to the storm, it spins horizontally, like a pencil rolling across the top of a desk. You can see the effect of this motion in the rolled look in the cloud of the storm's gust front.

As the air flow gets yanked up into the updraft, the angle of its spin changes. The pencil is still spinning, but it's been pulled up on its end, so it's spinning like a top.

If a supercell is going to produce tornadoes, this is where they often will appear — toward the rear part of the storm, in an area of twisted spin, where the rolling downdraft in the flanking line meets the rotating updraft.

A supercell is extremely dangerous (see Figures 9-9 and 9-10). Everything that is hazardous about thunderstorms is most hazardous about supercells, including their heavy rains and chances of flooding, the size of their hailstones, the strength of their winds, and the size and violence of the tornadoes they spawn. In this sense, they are more predictable than other thunderstorms. When forecasters identify a storm as a supercell, they *know* that the risks to public safety are high, and it's time to issue a severe weather warning.

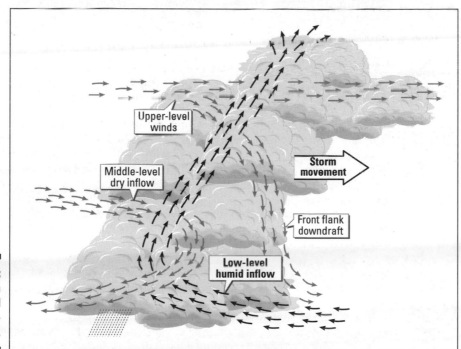

Figure 9-9:
Inside a supercell thunderstorm.

Figure 9-10:
A supercell thunder-
storm.

Jim Reed.

Hail the Size of Hailstones

It's pretty remarkable, when you think about it, these large pieces of ice falling from the sky in the warmth of a spring or summer day. Imagine the temperature differences and powerful wind shifts for such cold and heavy objects to form up in the clouds and then suddenly fall from the sky.

You would think that something that is responsible for more than $1 billion in property damages and crop losses ever year in the United States would get more respect. But no, when it comes to hailstones, they seem to exist only in relation to other objects.

Look what happens when people talk about hail: When it is relatively small, as it is most of the time, it is said to be pea-sized. When things really get cooking up there and it makes a few round trips of downdrafts and updrafts before falling an inch or so in diameter, then it is said to be golfball-sized. When it really gets out of hand, the sky lets loose with hailstones that are baseball-sized or softball-sized or even grapefruit-sized. Chapter 2 describes the formation of hail.

Hail of Fame

For the record, the largest hailstone in the United States measured 17.5 inches around and 5.5 inches across. That would be about the size of a softball, although there would be nothing soft about it. In my opinion, a hailstone that size deserves to stand on its own. Collected at Coffeyville, Kansas, on September 3, 1970, it weighed 1.67 pounds.

Hailstones are damaging and can cause injury, and once in a while, even death. In May 1986, an intense hailstorm in China killed 100 people and injured 9,000. Figure 9-11 shows you the average number of days of hail throughout the United States.

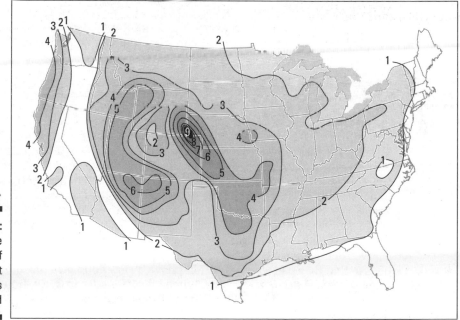

Figure 9-11: The average number of days that hail is observed

Flash Floods

Thunderstorms are the deadliest of all storms. Fast-moving thunderstorms bring more hail and tornadoes, but the slow-moving thunderstorms really are the worst. Their torrential rains bring a high risk of *flash flooding* — flooding that occurs in a matter of minutes or hours.

The National Weather Service calls flash floods "the Number One weather-related killer in the United States." Here's a comparison of the causes of the average yearly weather deaths across the country between 1972 and 1991:

- ✔ Hurricanes, 17
- ✔ Tornadoes, 69
- ✔ Lightning, 80
- ✔ Flooding, 146

Flooding is a lethal threat in every state. Chapter 10 describes floods in detail as well as the advice of experts about what you can do to prepare for these disasters.

ZAP! Crack! Bam!

Lightning is just too weird. Out of this tremendously wet and cold up-and-down turbulence of winds and ice crystals and rain comes this humungous *spark of electricity.* I mean, *are you sure?* I'm not much of an electrician, I have to admit, but still, up there in the sky in the middle of that blustery rain and hail — this really is *not* where I expect to find this stuff.

I'm not the only one. Lightning is one of those strange subjects that seems to get weirder and weirder the more you look into it. Scientists have this little phrase that they use when they don't have a very good grip on something. They say *it is not well understood.* When I hear those words, the tune that used to introduce the old television drama, *Twilight Zone,* pops into my head, and the hairs start to rise on the backs of my arms. (Do you suppose that it's the static electricity in the air?) Anyway, when they start describing the processes that produce the various electrical features of a thunderstorm, look for this little phrase and somewhere you should find it: *It is not well understood.*

For some reason, a giant cumulonimbus thunderstorm cloud, described in Chapter 5, develops positive and negative electrical charges in different parts. It's like a big battery. Check out Figure 9-12. The ice crystals on top become positive, and the water droplets and rain at the bottom become negative. This strong negative voltage at the bottom of the cloud has a way of pushing the negative voltage on the ground out to the side of the storm, because like-voltages repel one another. So the ground beneath the negative bottom of the cloud is made positive. These different electrical charges grow stronger and stronger, and then ZAP! — a spark, a lightning bolt.

The same thing happens in cloud-to-cloud lightning, which is the most common sort. The positively charged portion of the cloud, usually near the top, sparks with the negatively charged parts of another cloud, and the sky lights up with a giant streak.

Four out of five lightning bolts are within the cloud of a thunderstorm or flash from one cloud to another. About one in five travel from a cloud to the ground, which is enough already.

A stroke of lightning heats the air it strikes up to 54,000 degrees — roughly five times hotter than the surface of the Sun. Go figure.

Lightning is the most common cause of wildfires, a particular menace in the western United States. Each year, about 10,000 fires are started by lightning in the United States.

Lightning strikes, and the air explodes, and the sound of the shockwave is *thunder*. The great boom lets go at the same instant as the lightning, but the sound waves travel so much slower than the light waves that the thunder always seems to follow the lightning. Figure 9-12 shows the electrical charges and lightning that occur inside a thunderstorm

Can lightning strike the same place twice? You bet it can. In fact, when you think about it, whatever led lightning to strike the first time can lead it to strike there again. Unless it's no longer standing, of course. It is not uncommon for tall buildings to be struck often by lightning.

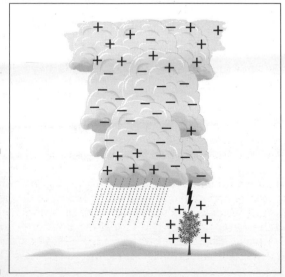

Figure 9-12:
The electrical charges and lightning inside a thunderstorm

Counting down to thunder

The speed of light: 186,000 miles per second.

The speed of sound: 1,100 feet per second.

For all practical purposes, at the speed of light you see lightning the instant it flashes. Hearing the thunder is another matter, and the huge difference between the travel times of these two waves allows you to estimate the distance to the thunderstorm.

It takes the sound of thunder five seconds to travel a mile. So if you count the seconds between the instant you see the lightning and the instant you hear its thunder, you can estimate how far you are from the lightning. If ten seconds passes, for example, it's two miles away.

Why does thunder seem to last so long while lightning is so brief? Again, it's the speed of sound. It takes much longer for the thunder to reach you from the lightning stroke at the top of the cloud than from the streak you see near the ground.

Rumbles and claps

The sound of distant lightning arrives as a deep rumble. It's a much deeper, more muffled and uneven sound than the enormous crack — or *thunderclap* — and explosive boom from lightning that strikes nearby.

Why so different? Along the way, the higher pitched cracking parts in the sound wave get absorbed or scattered, and that really deep, low pitch is all that survives the trip. The lightning strike itself can cover quite a distance across the sky, some of it closer and some of it farther away, and the difference in the distance accounts for that uneven rumbling. Some parts of the sound have farther to go than others, and so arrive later.

How far away can you hear thunder? When it comes to traveling sound waves, a lot depends on conditions in the atmosphere, and as you can imagine, when lightning is in the sky, conditions can get pretty goofy. On average, ten miles is about the limit, although as much as 25 miles may be possible under some rare circumstances.

There is a name for lightning that is too far away for its thunder to be heard. It's called *heat lightning*. This has nothing to do with the heat or the lightning, of course, but only your distance from the flash. At this distance, usually you see it reflected off the face of a cloud.

Different strokes

The lightning that you see as a streak or a fork is called *streak lightning* or *fork lightning*. If the clouds obscure the streak, it is called *sheet lightning*. But lightning also comes in other shapes and sizes and colors that you wouldn't believe.

Ball lightning

Once in a very great while, something called *ball lightning* is formed. A ball the size of a grapefruit or even as big as a basketball comes floating through the air, usually near the ground. It glows yellow or reddish, and it gives off a sizzling or crackling sound for several seconds before it disappears. This is truly an Unidentified Flying Object — a UFO. How this happens, nobody really knows.

St. Elmo's Fire

Another weird and rare thing happens sometimes in the middle of a thunderstorm. Tall objects like the mast of a ship or the steeple of a church will begin to glow and sizzle and hiss and spark and the air around it will give off a strange greenish glow. Nobody knows why.

Sprites and jets

In the last few years, even weirder things have been spotted at the very tops of some of the really big thunderclouds. Astronauts aboard the Space Shuttle have reported some interesting views of this. While lightning is flashing far below, up at the very tops of some clouds, a gigantic red blob flashes as much as 60 miles above the cloud. This bright reddish thing has strange blue-green tentacles dangling from it.

Also, there are mysterious blue jets squirting up from the tops of some of these clouds. They travel about 60 miles a second and reach upwards of 30 miles above the clouds. Sprites and jets and ball lightning are the subjects of some interesting research, but for the time being, it is not well understood.

Lightning safety tips

Lightning is the thunderstorm's most dangerous threat. On average, it kills 80 people every year just in the United States and injures another 300 persons. Most people hit by lightning are not killed, and two out of three fully recover.

By the way, there is no truth to the notion that people struck by lightning still carry an electrical charge. It is perfectly safe and vitally important to give them immediate first aid.

When storms of a thunderstorm gather, when lightning is seen or the crack of thunder is heard nearby, a few things are worth remembering.

- Take cover in a building, if possible. Inside, avoid electrical conductors like telephone wires, plumping pipes, and stoves.

- Never take shelter under a tree. If out in the open, avoid tall structures, metal objects, and water.

- The lightning threat does not pass when the rain ends. Often, it strikes its victims when it is not raining or just after it has stopped.

- Cars are relatively safe, as are airplanes, because even if they are struck, the lightning will travel around the metal body of the vehicle, rather than through its occupants.

Downbursts

The late Ted Fujita, the famous tornado researcher at the University of Chicago, discovered a powerful and dangerous feature of tornadoes while surveying the damage done by the worst bout of tornadoes ever to strike the United States — the super outbreak of 1974. Flying over West Virginia, Fujita noticed a particularly clear starburst pattern of wind damage down below.

This led him to realize that under some circumstances, downdrafts that carry cold air and precipitation down from the great heights of a supercell or squall line can hit the ground with powerful blasts of wind. When powerful downdraft winds reach the surface, they spread gusts of straight-flowing surface winds outward in all directions at speeds up to 165 miles per hour.

Lightning rods

When you hear it said that someone "acts as a lightning rod" on a particularly controversial subject, the expression very nearly reflects what Ben Franklin had in mind with his invention 250 years ago.

The phrase means that someone is standing out and attracting all the criticism and attention, deflecting it away from others, who are able to go about their business.

What Franklin invented was a metal rod that sticks up higher than any other point at the top of a church steeple or other tall building that is an easy target for a lightning strike. Attached to the rod is wire that runs the length of the building and extends down into the ground. The rod attracts the lightning, which otherwise would strike the church, and the wire conducts the charge harmlessly to the ground.

Fujita coined the terms *downburst* to describe the wind and decided they come in two sizes — *microbursts* that cover 2.5 miles or less, and larger *macrobursts.*

Downbursts can be invisible, clear air, or they can be accompanied by lashing rains. A dry downburst can fall out of *virga,* rain that evaporates before reaching the ground. The evaporation process cools the air, making it heavier than the air around it and causing it to speed downward.

What is peculiar about downbursts, however, is not so much their wind speeds as their abrupt change in wind direction. Such a wind formation poses a particular threat to airplanes that are landing or taking off at airports.

Put yourself in the pilot's seat for a minute: Coming in for a landing, she detects a powerful headwind and noses the plane upward to sail down through the extra lift. As the aircraft nears the runway, however, it passes through the center of the microburst, and in an instant the powerful headwind is a powerful tailwind. The plane suddenly loses altitude — and sometimes there is not enough altitude to lose.

During the 1970s and 1980s, microbursts were blamed for 27 airplane crashes in the United States that killed 491 people and injured another 206.

The discovery of microbursts led to the installation of millions of dollars worth of special radars at airports around the United States during the 1990s to help air traffic controllers warn pilots when local conditions are ripe for these dangerous winds.

Because *tornado* comes from the Spanish word for turn, sometimes the Spanish word *derecho* — straight or direct — is used as a colorful way to describe macrobursts, the big winds that can run out straight ahead of a squall line. Behind an ugly gust front of dust and soil, dangerous and damaging straight-line winds can come with the force of a hurricane — at an estimated high speed of 167 miles per hour, it has the force of an F3 tornado.

These wind gusts, which generally last ten minutes or less, have been known to blow over trees and wreck farm buildings across hundreds of miles of countryside. Derechos most often occur at night as a squall line of thunderstorms develops across the Great Plains and spawns a series of downbursts in front of a long string of storms.

The key to these downbursts is the winds that are blowing a few miles above the ground. When the winds up there are strong and their air is dry, conditions for downbursts are ripe. Precipitation evaporates in the dry flow, dramatically cooling the winds and causing them to plunge to the ground.

Really Twisted Winds

Tornadoes are nature's most violent storms. Nothing that the atmosphere can dish out is more destructive. They can sweep up anything that moves. They lift buildings from their foundations. They make a swirling cloud of violently flying debris. They are very dangerous to all living things, not only because of the sheer power of their winds, but the missiles of debris they create.

Wind measuring instruments are destroyed by tornadoes, although according to reliable estimates, their winds can exceed 250 miles per hour. Flying at those speeds, pieces of straw can penetrate wood. According to most scientists, the top wind speeds in the strongest tornadoes are about 280 miles per hour.

In an average year, 1,200 tornadoes are reported in the United States, far more than any other place in the world.

On average, tornadoes cause 80 deaths in the United States every year and 1,500 injuries, although averages don't mean very much when it comes to these storms. In 1998, for example, 130 people died in tornadoes in the United States, including 42 who were killed in an outbreak in central Florida and 34 who died in a single tornado in Birmingham, Alabama. Most human casualties are people in mobile homes and vehicles. The deadliest single tornado struck on March 18, 1925. In three and a half hours, it traveled 219 miles through Missouri, Illinois, and Indiana, killing 695 people.

Most tornadoes, nearly 90 percent, travel from the southwest to the northeast, although some follow quick-changing zigzag paths. Weak tornadoes, or decaying tornadoes, often have a thin ropelike appearance. The most violent tornadoes have a broad, dark, funnel-shape that extends from a dark wall cloud of a large thunderstorm.

The Super Outbreak of 1974

The worst tornado outbreak in history occurred in the United States April 3–4, 1974. A total of 148 twisters touched down in 13 states and Canada — Alabama, Georgia, Illinois, Indiana, Kentucky, Michigan, Mississippi, North Carolina, Ohio, South Carolina, Tennessee, Virginia, and West Virginia. It lasted 16 hours. A total of 330 people were killed, and 5,484 were injured.

The Super Outbreak was the most tornadoes in the most states, but it was not the deadliest tornado outbreak. That was the Tri-State Tornado of March 18, 1925, when 695 people were killed in Missouri, Illinois, and Indiana.

Property damage from the Super Outbreak was estimated at $600 milllion. Especially hard hit were the states of Alabama, Kentucky, and Ohio. The most damaging and deadly twister hit Xenia, Ohio. A tornado touched down southwest of Xenia and destroyed half of the town. The death toll was 34 in Xenia and damage to property totaled $100 million.

There have been reports of some tornadoes that practically stand still, hovering over a single field.

Others crawl along at 5 miles per hour. But the average tornado travels 35 miles per hour, and some have been clocked at more than 70 miles per hour. A tornado in 1917 traveled a record 293 miles. The average width of a tornado's path is about 140 yards, although some have been reported to be more than a mile wide.

Most tornadoes occur between 3 p.m. and 9 p.m., although they have been known to strike at all hours of the day or night. They usually last only about 15 minutes, usually moving constantly, spending only a matter of seconds in any single place, although some have been known to stay on the ground for hours.

Twisters over water are called *waterspouts*. Some of them are merely the result of land-formed tornadoes moving out over water, but most of them are not. They are a different kettle of fish. More often, a waterspout is a whirlwind that forms over warm water. The experts say that waterspouts tend to develop under rapidly growing cumulus clouds. Tornadoes, on the other hand, form under clouds that have already matured into giant clusters or supercells (see Figure 9-13).

Waterspouts are weaker and smaller than tornadoes, although their winds can reach 90 miles per hour and can damage boats. The warm ocean waters near the Florida Keys seem to make more waterspouts than anywhere else, although they also occur over large lakes such as the Great Lakes or even high mountain lakes such as Lake Tahoe in the Sierra Nevada. The Great Lakes waterspouts usually come in late autumn when a cold air mass moves over the warm water.

Figure 9-13: When tornadoes are likely to occur.

The chase is on

There are storm chasers, and then there are the rest of us. If the very idea of chasing a severe thunderstorm so that you can get as close as possible to a tornado — the most violent and dangerous storm on the planet — sounds absolutely crazy to you, well *hold that thought*. It is the wiser part of human nature to seek shelter from the storm.

Besides, by all accounts, there are already enough people out there driving along those narrow country roads of the Plains, through the winds and the rains, stalking the wily tornado. Some are meteorologists with mobile Doppler radars doing serious science. Some are out there more for the thrill of witnessing the incredible fury of nature.

Many storm chasers do valuable work, and their daring has become an important part of tornado science. The mobile radars are seeing ever more clearly the fine details of air flows inside of storms. Observations continue to add new wrinkles and features to ideas about how and why one storm forms a tornado and another storm doesn't. And the spotters provide valuable information to storm forecasters.

For the rest of the world, there are the videos they bring back that end up on television. And, of course, in the warmth and safety of our homes now, there is the movie "Twister."

Tornado Alley

The size of the place known as Tornado Alley expands through spring and summer as heating from the sun grows warmer and the flow of warm moisture from the Gulf of Mexico spreads farther north. An area that includes central Texas, Oklahoma, and Kansas is the hard core of the season, but before it is over, as Figure 9-14 illustrates, Tornado Alley extends north to Nebraska and Iowa.

It shrinks and swells over time, but there is only one Tornado Alley. Nowhere else in the world sees weather conditions in a combination that is so perfect for these storms. Here's what makes the storms of Tornado Alley so bad:

  ✔ Beginning in spring and continuing through summer, low-level winds from the south and southeast bring a plentiful supply of warm tropical moisture up from the Gulf of Mexico into the Great Plains.

✔ From down off of the eastern slopes of the Rocky Mountains or from out of the deserts of northern Mexico come other flows of very dry air that travel about 3,000 feet above the ground.

✔ From 10,000 feet, the prevailing westerly winds, sometimes accompanied by a powerful jet stream, race overhead, carrying cool air from the Pacific Ocean.

Sometimes, the winds form a convective cap lid of warm air over the Plains that the rising air is eventually able to break through and explode upward into the sky (see Figure 9-15). These are the ingredients for the most severe thunderstorms and most powerful twisters — sharp differences in temperatures at different levels, big contrasts in dryness and moisture, and layers of powerful winds that are blowing from different directions at different speeds.

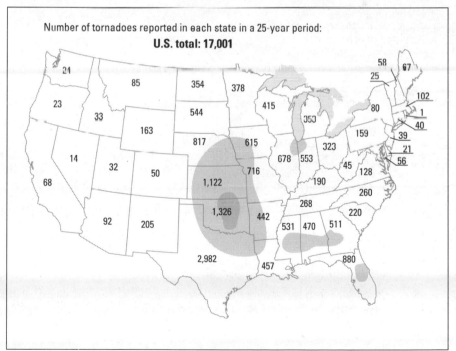

Figure 9-14:
Tornado
Alley.

Figure 9-15:
A thunder-
storm
explodes
upward.

National Oceanic and Atmospheric Administration/National Climatic Data Center.

Forecasting

Weather forecasters in Tornado Alley have a pretty good idea of the menu of conditions that are necessary to make severe thunderstorms, and they're pretty good at being able to forecast that severe thunderstorms are on the way. They can say that large hailstones and strong winds are likely, and a tornado is a possibility during the next several hours or the next day or two.

But they can't forecast a tornado. The question of which of the conditions on the menu for severe thunderstorms actually causes tornadoes to form in these storms remains one of the most difficult mysteries of weather science. A severe thunderstorm that causes a tornado can look exactly like a severe thunderstorm that does not cause a tornado. Weather researchers have been working on the problem for years, chasing tornadoes all over the country-side, and still it is one of those things that is not well understood.

The presence in the area of supercell thunderstorms really puts pressure on forecasters in local weather service field offices. The national Storm Prediction Center in Norman, Oklahoma, is on the phone giving advice, but the buck stops in the local office. The local forecasters know that a lethal tornado could come spinning down out of the dark cloud at any moment, but they can't be sure until they see it show up on a Doppler radar screen or a funnel is actually observed.

Billions of dollars have been spent in the last several years on research and computer modeling and radars and satellite technologies and high-speed communications, and progress has been made. On average, when tornado warnings were issued in 1994, communities had six minutes to react. By 1998, the average lead time for warnings had been stretched to 12 minutes.

Television meteorologists and other media outlets play vital roles in such weather emergencies, continuously broadcasting the locations and predicted paths of tornadoes. Many lives are being saved by the increased public awareness and the lengthening time of advance warning that is available. In fact, the longer lead-time has reached the point where people are rethinking the idea of public shelters for tornadoes. As minutes are added to advance warnings, now it may be possible for people in harm's way to rush to a shelter before a tornado hits.

More than 15,000 severe storm and tornado watches and warnings are issued by the National Weather Service every year. Most of the time they are accurate. Sometimes they are missed. Occasionally there are false alarms. The successes are taken for granted and often overlooked in the details of a tornado disaster. The failures and the false alarms seem to be remembered forever. Perfectly reasonable people who will forgive you for missing the rain on their picnic now have a different attitude. When it comes to tornadoes, they want perfection.

A famous tornado researcher, the late Ted Fujita, at the University of Chicago, devised a scale to measure the power and damage of these whirlwinds. When meteorologists and storm-chasers talk about Category-one or F-3 tornadoes, they are using the Fujita Scale.

Table 9-1 shows you how to use the Fujita Scale to tell one tornado from another.

Finlay's secret storms

The world's first tornado expert was a U.S. Army sergeant named John Finlay, who was assigned the job of researching these storms in 1882. The National Weather Service was a brand new agency under the army's Signal Service, and Finlay was to figure out how to predict these mysterious killer storms.

By 1887, he had a network of 2,400 volunteers reporting tornadoes, and Finlay's pioneering research was leading to better understanding of what it takes to make a twister. He was focusing his early prediction efforts on dew-point and winds.

Then a curious thing happened. The Signal Service officials decided that public panic was a bigger threat than tornadoes. At a time when much of the western Plains were still being settled, did it occur to the powers-that-be that Finlay's results were not good for real estate values? In any event, they banned the use of the word "tornado" in forecasts and abolished Finlay's research unit.

"Tornado" was officially forbidden until 1938, and it wasn't until the 1950s that the nation's weather service began issuing forecasts for severe thunderstorms and tornadoes.

Table 9-1	Interpreting the Fujita Scale
Category	*Description*
Category F0	Gale tornado, winds 40–72 mph. Light damage. Some damage to chimneys, branches broken off trees, shallow-rooted trees pushed over, damage to sign boards.
Category F1	Moderate tornado, winds 73–112 mph. Moderate damage. Roof surfaces peeled off, mobile homes pushed off foundations or overturned, moving cars pushed off the roads.
Category F2	Significant tornado, winds 113–157 mph. Considerable damage. Roofs torn off frame houses, mobile homes demolished, rail cars pushed over, large trees snapped or uprooted, light-object "missiles" generated.
Category F3	Severe tornado, winds 158–206 mph. Severe damage. Roofs and some walls torn off well-constructed houses, trains overturned, most trees in forest uprooted, heavy cars lifted off the ground and thrown.
Category F4	Devastating tornado, winds 207–260 mph. Devastating damage. Well-constructed houses leveled, structure with weak foundation blown off some distance, cars thrown, and large missiles generated.
Category F5	Incredible tornado, winds 261–318 mph. Incredible damage. Strong frame houses lifted off foundations and carried considerable distance to disintegrate, automobile-sized missiles fly more than 100 yards, bark torn from trees, incredible phenomena occur.

Lives and Limbs

What are the odds of a tornado crossing your path? Even in Tornado Alley, the odds are against such an unhappy occasion. When it happens, of course, it's a disaster — but still, the odds are high against it.

People think about tornadoes in tornado country the way people in the Southeast think about hurricanes and people in California think about earthquakes. It's part of the background of daily life that you really don't give very much thought to, because chances are, it's not going to happen.

Tornado myths

For a long time, people were told to open the windows on the sides of their houses that are away from the path of an approaching twister. This was to equalize the air pressure inside and outside the house, because experts thought that the tornado's low pressure caused the house to "explode."

Well, it turns out that houses don't really "explode," the debris pile just looks like it sometimes. Often, the wind lifts the roof off, and the walls are peeled away like onion skins. Also, weather scientists realized that the most powerful storms on Earth were not going to be affected one way or another by a few panes of glass.

There are a number of myths about tornadoes, such as the thought that hills protect areas from tornadoes, and that mammatus clouds, which hang from the undersides of thunderstorm anvils, spin out tornadoes.

The Super Outbreak of 1974 debunked several myths in just a few fell swoops.

One was that a tornado won't touch down where two big rivers come together. On April 3, 1974, a tornado hit the town of Cairo, Illinois, where the Ohio River joins the Mississippi.

Another myth was that tornadoes won't go up or down hills. One Super Outbreak tornado hit Guin, Alabama, climbed 1,640-foot Monte Sano Mountain, and roared down the other side. Another formed at 1,800 feet east of Mulberry Gap, Tennessee, crossed a 3,000-foot ridge, moved down the canyon, and then climbed 3,300-foot Rich Nob.

Do tornadoes stay out of steep valleys? That was another myth that bit the dust April 3, 1974. A tornado hit Monticello, Indiana, wiping out three schools, and then descended a 60-foot bluff over the Tippecanoe River and damaged homes.

The five-dollar word for this is *complacency* — a self-satisfied unawareness of danger — and somebody is always getting on their high horse about it. The truth is, day in and day out, most people have other things to worry about that just seem more real. And it's just human nature to be optimistic, and to think things are going to turn out for the best. But it leaves you and I open for some terrible surprises once in awhile, which is kind of sad, when you think about it. Government people in the disaster business and American Red Cross relief workers who deal with victims of these storms see this sense of surprise on people's faces all the time.

A watch or a warning?

Don't confuse a "watch" with a "warning." There is a big difference. Here is what they are about:

✔ **Tornado Watch:** When National Weather Service forecasters issue a Tornado Watch, they are making *a forecast that tornadoes are possible* in your area. It's time to remain alert to signs of approaching storms and to make sure that you are prepared for an emergency.

✔ **Tornado Warning:** This is an *emergency message*. A tornado has been sighted in your area, or weather radar indicates one is present. Now is the time to get to safety, to put your emergency plan into action.

Tornado do's — and nots!

The National Weather Service and the American Red Cross have put together these basic tips about tornado safety:

✔ Seek shelter immediately, preferably underground in a basement, or in an interior room on the lowest floor such as a closet or bathroom.

✔ Stay away from windows

✔ Get out of your car or your mobile home and seek shelter in a sturdy structure. In the open, lie flat in a ditch or depression.

✔ Protect your head from flying debris.

✔ Do not try to outrun a tornado in your car.

✔ Do not seek shelter under a bridge over overpass. The idea that these are safe shelters is just plain wrong.

The fastest wind on Earth

One of the most powerful tornadoes ever observed struck the outskirts of Oklahoma City on May 3, 1999. This giant twister, a half-mile wide, had the fastest winds ever recorded on Earth. Radar readings of the F-5 tornado that hit Moore, Oklahoma, that day measured the winds at 318 miles per hour. The fastest winds were between 150 feet and 300 feet above the ground. The path of this tornado covered more than 40 miles.

The twister was one of 78 tornadoes that formed the afternoon and evening of that day in Oklahoma, northwest Texas and south-central Kansas. The outbreak killed 48 people and injured 800. Most of the casualties were in Oklahoma, where 1,800 homes were destroyed, another 2,500 damaged.

The twister that hit Moore that day set other records besides wind speeds. It damaged almost three times as many structures as any previous American tornado and caused property losses that were estimated at more than $1 billion. It was the nation's first billion-dollar tornado.

Chapter 10

Extremely Summer

● ●

● ●

*L*et the poets fawn over spring, the April showers and the May flowers and what-have-you. Any school kid will tell you that summer is the season that weather is all about. All of those long, dark days of winter and the dampness of spring are things you have to put up with just to get the world ready for another good dose of summertime.

Sure, it rains in the summer over the eastern two-thirds of the United States, but most days it comes and goes. The rainstorms of summer usually are not like the huge, cold, and powerful dead-of-winter storms that can keep a kid trapped indoors all day.

But summer can be a season of extremes. It is the time of year in the region of the middle latitudes of the Northern Hemisphere, where you and I live, when the Sun's energy is having its greatest effect on the atmosphere, the land, and the sea. When the Sun's heat energy in the atmosphere is at its peak, a lot can go wrong in a hurry. If conditions are unusual in the atmosphere or in the ocean or over the land, the weather that results also can be unusual in a very big way.

The good ol' summertime can be a time of weather disasters. Summer's warmth suddenly can turn into a staggering heat wave. It is a time when severe thunderstorms are possible over a large portion of the eastern two-thirds of the United States. The threat of tornadoes, which Chapter 9 describes, is at its peak. A single thunderstorm can cause a local flash flood. A large, stubborn pattern of thunderstorms can bring days of torrential rains that cause big rivers to spill out of their banks.

At the other extreme, summer is the time when drought — the lack of rainfall — is most likely to get everybody's attention. The stage may have been set months earlier, when winter rain or snow failed to arrive to replenish reservoirs or groundwater. But now, in the parched heat of summer, is when the water is most sorely missed. The dangerous summer extremes of heat and rain and drought is what this chapter is all about. Chapter 7 describes tropical storms and hurricanes, which also begin showing up along the south and eastern coasts this time of year.

Good Ol' Summer Timing

As with all seasons, which are controlled by Earth's tilt as it orbits the Sun, which Chapter 2 describes, summer means something to astronomers that doesn't have anything to do with the weather. Ask an astronomer, and she will tell you that, rain or shine, summer begins on June 21 in the northern half of the world. That is the day of the *summer solstice* in the Northern Hemisphere (and the *winter solstice* in the Southern Hemisphere). As Figure 10-1 illustrates, June 21 is the day when Earth reaches the place along its path in space when its tilt allows the highest amount of the Sun's radiation to reach the Northern Hemisphere.

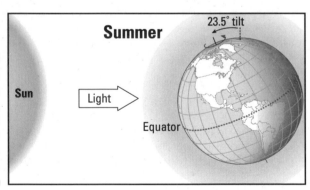

Figure 10-1:
Sun is at highest point in the sky over Northern Hemisphere.

Looking at the situation June 21 from Earth, the daily path of the Sun across the sky has reached its highest point overhead, and from now on, the hours of daylight will become shorter. At noon that day, the Sun's track has reached as far north — as high overhead — as it is going to go. It will seem to sort of stay on that same track for a few days before you notice that it has started back toward the south. This impression that the Sun has stopped in its tracks is where the old Latin word *solstice* comes from: It means "sun stands still."

June 21 does not make the same impression on somebody looking at the Northern Hemisphere's sky as it does on somebody with stars in their eyes. On one hand, if you think of it as the first day of summer, like an astronomer, in most places in the United States, in most years, June 21 feels pretty late. Usually, the weather of summer has been around for awhile. On the other hand, if you think of it as the beginning of the end of summer, when the Sun's light is starting to fade from the Northern Hemisphere, June 21 feels too early.

Even though the sunlight starts to fade after that day, the temperatures in the Northern Hemisphere don't really start cooling off during the days just after June 21. In fact, in the middle latitudes where you and I live, the warmest weather of the summer usually comes several weeks later, in July or August. Weather scientists explain the difference in timing this way: Even though daylight hours are growing shorter after June 21, for several weeks the amount of heat energy arriving from the Sun will still be more than the amount that the warmed Earth radiates back into space. So the summer warmth continues to build up. This *seasonal lag* is the time that it takes for Earth's atmosphere and land and oceans to respond to these sunlight changes.

The seasons come and go like clockwork in the eyes of astronomers. In the Northern Hemisphere, they will tell you, summer lasts 93.65 days. But the weather of summer doesn't work that way. The Sun rises on time, it's true, and the things that combine to make the season's heat waves and storms and floods and droughts all work together. But as every school kid and weather forecaster can tell you, nothing important about summer reminds you of a clock!

Coast to Coast

Summer is all about heat. The Sun has won back the land that it gave up during autumn and winter to the cold air masses of North America. Once again, it's the star of the show. The months of summer are generally considered June, July, and August, but the Sun's energy — measured by the average daily temperature — really is the long and short of it. When average daily temperatures hit 68 degrees, summer has arrived.

As Figure 10-2 illustrates, in the eastern two-thirds of the United States, when your summer arrives and how long it stays depends most of all on your *latitude* — the distance between you and the Equator. That's the long and short of it. If 68-degree daily average temperature is the general rule, summer arrives as early as March 1 in southern Florida and as late as July 1 in northern Maine.

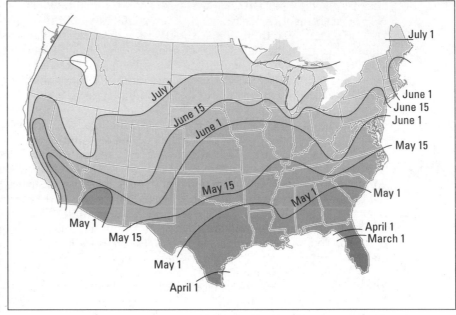

Out West it's a different story. It is still true that latitude is important, of course. The desert climates of the Southwest are as famous as Florida for their long and hot summers as well as their short and warm winters. And the differences between the Southwestern summer and the Northwestern summer certainly are big. But two other large features in the western third of the United States have a lot to do with summer temperatures and weather. While the valleys get hot in summer, in large areas of the mountainous west — the Rocky Mountains, the Sierra, and the Cascade ranges — daytime temperatures stay cooler because of their elevation. Second, the prevailing westerly winds keep flowing over much of the west. Pacific Ocean air is drier and more stable than flows prevailing elsewhere in the country.

Chilled by the Pacific Ocean's cold, southward-flowing California Current, the air flowing inland over northern California and much of the west has a *dewpoint* — or temperature to reach saturation — of 50 degrees or so. This is nearly 20 degrees lower than the dewpoint of air near the Gulf Coast and over the Southeast. Flowing inland, and heating up over the western deserts, this Pacific air can reach a bone-dry relative humidity of 10 or 20 percent. It rises in this heat, of course, but it has a long way to go to reach its saturation point and create clouds. At the water's surface, evaporation is putting a lot more water vapor into the air above the Southeast's warm Gulf Stream than the West Coast's cold California Current.

The clues to the biggest differences between summer weather patterns in the eastern two-thirds of the country and summer weather out West are out in the oceans. As the Sun's energy warms the Northern Hemisphere, two large high pressure areas grow bigger and bigger out at sea — the Pacific High off the West Coast, and the Bermuda High off the East and Gulf Coasts. You may think that these two high pressure systems would have the same effect on weather in the United States, but just the opposite is true.

As Figure 10-3 illustrates, the most important difference is this: The West Coast is to the *east* of the Pacific High, and the East Coast is to the *west* of the Bermuda High. Because winds circulate clockwise around high pressure in the Northern Hemisphere, the two coasts get flows of air from very different parts of their neighboring oceans. Out West, the northerly flow keeps summertime conditions drier and cooler than they would be otherwise during the summer. In the East, the southerly flow of the neighboring Gulf and Atlantic makes some days more humid.

The vegetation on the landscape contributes to these big differences as well. In the West, the parched ground produces a plant life that survives by retaining moisture through the long, dry summers in the region. In the East, humid conditions produce a widespread leafy vegetation that feeds back large amounts of moisture into the air.

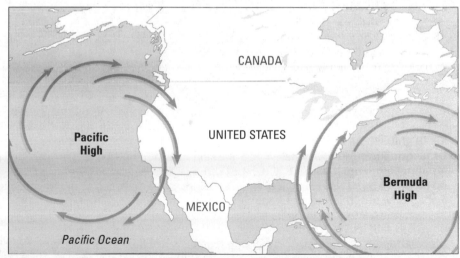

Figure 10-3:
Different coasts, very different air flows.

These big air flow differences affect more than temperatures. They mean that in most of the western third of the United States, summer is a dry season. Outside of the rainy Pacific Northwest, practically all of the rainfall arrives in the months of winter. In the East, that constant flow of tropical moisture means that it can rain all summer long. Check out Figure 10-4 and see the difference in the rainy seasons between Sacramento, California, and Washington, D.C., two cities that are on opposite sides of the country at pretty much the same latitude.

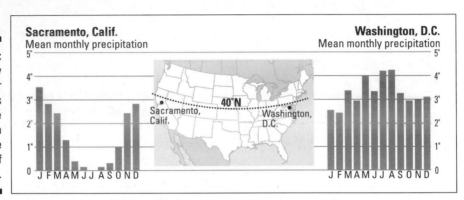

Figure 10-4: Rainy seasons for two cities at same latitude on opposite sides of the U.S.

Avoiding That Radiant Feeling

Among people who know the effects of exposure to too much solar radiation, worshipping the summer Sun has lost a lot of its charms, even in places like sunny California.

It's not just the immediate pain of sunburn, although that can be bad enough. Especially if your skin is light, your complexion fair, overexposure to sunlight can come back to haunt you years later in more serious forms of suffering.

Sunburn is skin damage. Your skin has the ability to repair this damage, but only up to a point. The more often and more seriously you subject your skin cells to the damage of overexposure to sunlight, as a rule, the sooner you reach the limits of your skin's ability to repair the damage, and the more serious the consequences.

Sun damage can lead to skin cancer years later. Skin cancer is the most common form of cancer. About 1 million cases of skin cancer are diagnosed every year in the United States. Most skin cancer, but not all of it, can be easily treated if detected early. The disease kills more than 9,000 people a year in the United States.

The UV Index

Here are the official National Weather Service Ultraviolet Index exposure levels and protective measures suggested by the U.S. Environmental Protection Agency:

Index	UV Exposure	What To Do
0 to 2	Minimal	Apply skin protection factor (SPF) 15 sunscreen
3 to 4	Low	SPF 15 and protective clothing (hat)
5 to 6	Moderate	SPF 15, protective clothing, and sunglasses
7 to 9	High	SPF 15, protective clothing, sunglasses. Try to avoid the Sun between 10 a.m. to 4 p.m.
10 or higher	Very High	SPF 15, protective clothing, sunglasses, and avoid being in the Sun between 10 a.m. to 4 p.m.

Scientists have isolated the dangerous part of the sunlight. It's the ultraviolet rays that travel at a wavelength of light that you and I can't see. A layer of ozone high in the atmosphere filters out a lot of the UV radiation, but not all of it, and during summer, when the Sun is highest in the sky, levels of UV radiation can be especially hazardous.

The National Weather Service and other forecasting services have begun issuing a daily UV Index Forecast for many cities around the United States. Weather scientists measure the ozone layer and other features like cloudiness and predict the intensity of the ultraviolet radiation reaching within about 30 miles of the city. It ranges from 0 to 15, from low to high UV intensity.

The Heat Is On

Wet or dry, summers in the United States are hot. Blame this great big North American continent that you and I live on. The same thing that causes those giant masses of cold air to form in winter is responsible for big masses of hot air in the summer. Land reacts to the coming and going of the Sun's heat energy in a big way, much more so than water. It absorbs the heat more easily and loses it quicker. The bigger the continent, as a rule, the more common the extremes of temperature. Most summers in the United States see dangerous heat waves in one place or another.

The shadow knows

So why are temperatures always measured in the shade? After all, when I get really hot, as often as not it's because I *haven't* been in the shade, I've been too long in the Sun. Why let the thermometer cool it all day in the shade and tell me what the temperature around me would have been if I had?

There's more than one good reason for measuring temperatures in the shade:

✔ A thermometer is made of material — glass, plastic, metal — that would warm very quickly and make the instrument's readings radically higher than the air around it. In no time, you would have a lot of broken temperature records — and a lot of broken thermometers.

✔ Depending on such conditions of cloudiness and air quality, the intensity of sunshine goes up and down more often and more quickly than the temperature of the air. And the air can warm up or cool off for reasons that don't have anything to do with the sunshine overhead.

✔ Taking temperatures in the shade is a standard that allows measurements to be compared the same way all over the world. Everybody's sunlight might be a little different from everybody else's from one day to the next, but everybody's shade is the same.

West of the Rockies, it's a *dry heat,* as the saying goes, meaning it is not very humid — but boy, can it get hot! The Pacific shoreline is bathed in cooling ocean breezes. In fact, San Francisco often is under a fog bank during summer. But travel inland and the temperature rises mile by mile. Leave San Francisco in 65 degrees some summer afternoons, drive 90 miles down the road, and your poor body doesn't know what to do with itself — in the Central Valley, it's 105 degrees in the shade. (Incidentally, temperatures are *always* measured in the shade.) And in the desert Southwest, high pressure is building, and the place is ablaze this time of year. In Yuma, Arizona, for example, 107 degrees in July is normal.

East of the Rockies, some of the time it's a *wet heat,* as high temperatures combine with high humidity, but conditions are more changeable. A westerly wind can carry air that is so hot it breaks temperature records, but it is relatively dry air. A north wind can bring dry but relatively cool air from the north. Coming from the south or southeast, winds bringing in tropical warmth from the Gulf of Mexico or Atlantic can give the region a real bath.

Also, the warm temperatures of summer are more evenly spread around than the cold temperatures of winter. There is a bigger difference in the intensity of sunshine between north and south in the winter than the summer. The mid-latitude storms, which feed off the differences between warm and cold air masses, are strong in winter and weaker in summer.

In January, most often there is a big difference between temperatures in Houston, Texas, for example, and Minneapolis, Minnesota. In summer, however, most of the time a big, warm blanket of air covers the entire region, and the temperature differences are not as great. Sometimes, when the northward flow of warm moisture from the Gulf of Mexico is strong, there is little difference in temperatures and humidity between the southern Plains and the upper Midwest.

Heatwaves

When summer temperatures are on the rise, weather forecasters begin to worry when high pressure squats over a region for too long. They are keeping their eyes peeled not just on the temperatures you are feeling on the ground, but also on the temperatures in the air over your head. They are looking at the temperature *profile* of the air to see whether it continues to get cooler the higher it goes.

If the air is warmer overhead than it is at the surface, the building heat will be prevented from escaping up into the higher altitudes. These conditions form a temperature inversion that acts as a lid over the region, preventing the normal process of nighttime cooling. Not only does the heat stay low to the ground, but so does the air pollution. It's like living in a hot room without any ventilation.

Chicago's summer of death

Like everybody in the region, people in Chicago had heard that a heatwave was in the forecast. But hot, muggy weather comes with the season in the Midwest, and it's something that everybody knows they just have to put up with once in awhile. Nobody could foresee what would hit the city in four days in July 1995. It had never happened before.

The heat grew intense on July 12, climbing well above the 90-degree mark, and an inversion layer of warm air overhead trapped the heat near the ground like a lid on a pot. But it was not just the heat. The spring had been especially wet, and humidity was very high. The dewpoint rose that night to 76 degrees, meaning that the trapped air was so saturated that it would not absorb any more water vapor through the cooling evaporation of sweat from the body. This also had the effect of keeping night-time temperatures especially high for two days.

Temperatures reached 104 degrees on July 13, and the high humidity made it feel like 119 degrees.

For four days the city sweltered, and by the end of it more than 700 people were dead from the effects of the heat. Many of the victims were elderly, and many were poor, without access to air conditioning. Many lived alone, and many lived in the top floors of apartment buildings.

Chicago's tragedy changed the way public health officials and emergency managers think about and prepare for heatwaves. Now officials in large cities are on the lookout for these bouts of high temperatures and high humidity. They have plans to open air-conditioned public buildings and alert citizens to the threat. And the National Weather Service has devised computer models that give up to two days' advance warning of a coming heatwave.

How heat kills

Your body is always generating heat. It wants to keep a certain level of heat inside — 98.6 degrees — but not more, so it needs a cooling system. Your heart, your blood vessels, your skin, and the glands that make you sweat all work together to do this job. Tiny blood vessels carry warm blood into your skin, which radiates it away. When necessary, glands pour sweat out onto your skin, and loss of the heat that is used to evaporate the liquid cools you off more.

Everything works pretty well until the temperature outside of your body rises above 90 degrees and humidity is high. When the air already contains a lot of moisture — as it approaches its *dewpoint*, in other words — it doesn't easily take up any more water vapor through evaporation. Evaporation rates depend on temperature, so your hot, perspiring skin is shedding plenty of moisture. Trouble is, when it is very hot, water vapor molecules are jumping onto your skin as fast or faster than evaporation can send them away. So even though you keep sweating, you don't really get any cooler. Now your heart

is pumping extra hard, although you may not notice it, but your cooling system is on the blink. A lot of warm blood is flowing through your skin, but the temperature outside your body is almost as warm as it is inside your body. And your glands are pouring out liquid for the air to drink up and cool you off, but the air isn't thirsty.

The temperature of your body's inner core begins to rise above 98.6 degrees, and you are flirting with serious illness. You're sweating heavily. Painful muscle spasms called *heat cramps* may develop in your legs or abdomen. As your body temperature continues rising, you grow weak, pale, and clammy. Your pulse grows weak. You faint and vomit. Still the heat continues climbing inside your body. When your body temperature reaches 106 degrees, instead of sweating, your skin is hot and dry. Your heart is pumping frantically, and your pulse is rapid and strong. You lose conciousness.

Now, is this any way to treat a body?

The National Weather Service defines a *heatwave* as a period of two days in a row when apparent temperatures on its official Heat Index are likely to be greater than 105 degrees to 110 degrees, although these standards depend greatly on local climate. A heatwave in Tucson, Arizona, does not kick in at the same temperatures as a heatwave in Duluth, Minnesota.

Chapter 1 describes the official Heat Index that combines the effects of high temperatures and humidity. This idea is like the *wind chill factor* that let's you know how cold it really feels when the wind is blowing at one speed or another. The Heat Index gives you the *apparent temperature* — how hot it really feels to the body in the shade when humidity is added to the actual temperature.

Heatwaves are very dangerous, and forecasters' warnings about them should not be ignored. They are not just matters of discomfort or inconvenience. Your health and even your life can be at risk. The National Weather Service reports that in a normal year about 175 Americans die from the effects of the summer heat. This is a greater death toll than lightning, hurricanes, tornadoes, or floods. According to another estimate, every summer health officials in the United States point to heat as the key cause of about 2,000 deaths. Whatever the numbers, on average, the heat of summer kills more people than all weather hazards, with the possible exception of the cold of winter.

One study found that in the 40-year period from 1936 through 1975, nearly 20,000 people were killed in the United States by the effects of heat and the Sun's radiation. In a disastrous heat wave of 1980, more than 1,250 people died.

And these are the direct casualties. How many more deaths are advanced by heatwave weather? How many diseased or aging hearts give out that might have continued working under better conditions? Nobody really knows, but there are experts who believe that summer heat is the biggest killer of all.

As Heat kills by taxing the human body beyond its abilities. The elderly are especially vulnerable because the cooling mechanisms of their bodies do not work as well. The same conditions that might cause heat cramps in a 17-year-old might cause more serious heat exhaustion in a 40-year-old, and in a person over 60 it can mean dangerous heat stroke.

Stranded on crowded islands

Cities pose special hazards during periods of especially high summer temperatures. Weather scientists refer to it as the *heat island effect*. The Sun's rays beat down on pavements and sidewalks, the roofs and sides of buildings, and these surfaces absorb and hold the heat and are slow to give it up, especially at night. Day and night temperatures in a city can be several degrees hotter than the surrounding countryside.

Cooling it

Experts at the National Weather Service and the American Red Cross offer these personal safety tips for people caught in a heatwave:

✔ **Slow down.** Avoid strenuous activity. Your body is already working extra hard.

✔ **Dress lightly.** Lightweight, light-colored clothes reflect heat and sunlight and keep you cooler.

✔ **Eat lightly.** Foods like proteins increase body heat and water loss.

✔ **Drink heavily.** Your body needs plenty of water to keep cool, but avoid drinking alcohol, which dries out the body.

✔ **Stay out of the Sun.** Especially avoid sunburn, which makes it hard for the skin to release body heat.

✔ **Find air-conditioning.** If you do not have air-conditioning where you live, spend as much time as possible in an air-conditioned building, such as a library, theater, or other public place, such as a mall.

The calm, stagnant air also causes the continual buildup of hazardous pollution from toxic exhausts of vehicles and factories. Making matters even worse, constantly running air-conditioners means that demand for electricity skyrockets during heatwaves. Electrical power plants are usually located at remote distances from cities, and the long transmission lines also are stressed by the heat, sometimes leading to power failures.

Storms of Summer

The warm temperatures of summer have a large and long-term feel about them in many parts of the country. Sometimes you can imagine those long summer days lasting forever.

The storms of the season, on the other hand, look local and feel temporary. They are summer showers. Most of them pop up in the warmth of the late afternoon across much of the Midwest, although in the Plains, thunderstorms are common in the hours before dawn and after dusk. In a matter of minutes, they are gone, as quickly as they came. In a rush, first in big, blotchy drops and then in a sudden downpour, a little rain has fallen. The dust has been settled out of the air. Sometimes there is a strong scent of ozone gas that formed in the thunderstorm's lightning.

But a storm of summer is not as local or as temporary as it looks. Sure, it came and went like a flash in the pan — charging up and blowing off and then vanishing from the sky. But a weather scientist will tell you that the storm has left behind temperature and humidity conditions in the atmosphere that make it easier for another thunderstorm to take shape there the next day.

Sure, most thunderstorms are small and brief weather events, but the conditions that make them likely often are large features of the sky that can hang around for months. The summer's thunderstorms are creatures that thrive on heat and moisture. Take those features out of the summer sky and poof! — no summer rainfall. As Figure 10-5 points out, the flow that supplies the tropical moisture that fuels summer storms east of the Rockies is part of a very large and long-lasting feature in the atmosphere over the Atlantic Ocean. While individual storms are usually small and seemingly minor, they are really part of a big, season-long rainfall pattern.

Thunderstorms are sensational. At a safe distance, they might strike you as majestic creatures with their bright billowing castles of cumulus cloud and their great fibrous crowns spreading far and wide. Their lightning and thunder is always impressive. When they are on top of you, they can be fearsome events. Flash flooding is possible from their downpours. Their hail can destroy a crop in minutes. Their wind gusts can snap the branches of trees and twist their trunks. And their lightning is as dangerous as it looks.

And when conditions combine to allow thunderstorms to organize themselves into the larger systems described in Chapter 9, there is nothing brief or temporary about them. Their slow movement and torrential rains can bring flash floods that are some of the most dangerous weather disasters. They can even persist so long during some summers that very large river flooding disasters spread across the land.

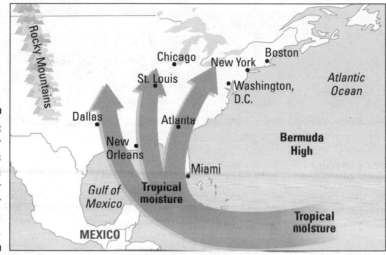

Figure 10-5: Summer moisture is part of a giant air flow over the Atlantic Ocean.

Out of Whack

When it comes to natural weather disasters in the United States and around the world, nothing in history compares to the devastation of the floods that follow too much rain, and the droughts and famines that follow too little.

Droughts and floods that result from season-long patterns of rainfall are related in an important way. If you're not getting the summer rain you are accustomed to, it probably is falling on somebody else who is not expecting it and doesn't want it nearly as much. Both *your drought* and *their flood* occur when important features of the ocean-atmosphere system are thrown out of whack.

Stubborn high pressure systems can take shape over places that usually experience low pressure, and low pressure systems are found in unusual places. Highs can block the normal flow of prevailing westerly winds in the upper atmosphere that help summer storms develop. Unusually shaped jet streams can push winter storms over deserts instead of forests.

Wrong turns

Experts say that nearly half of people who die during flash floods are people in cars or trucks whose drivers make two dangerous mistakes. Here's where they make the wrong turns:

- They miscalculate the depth of water crossing a street or highway, so they don't realize how deep it is.

- They underestimate the power of the water to pick up their vehicle and float it away.

Water is heavy. A gallon weighs eight and a half pounds, a cubic foot weighs 62.4 pounds, and a big bathtub of it weighs three quarters of a ton. Flowing water is powerful. Just six inches of fast-flowing water can knock you off your feet. If your car stalls in just a foot of flowing water, it typically has 500 pounds of force pushing against it. And in two feet of water, the average car floats away.

Chapter 6 describes the coupled ocean-atmosphere system and the climate conditions such as El Niño and La Niña that can lead to floods and droughts. Big changes in the temperatures of water at the surface of the oceans that last for several months are the kinds of things that can shift the locations of areas of heavy rainfall and dryness around the world.

For example, La Niña conditions, when temperatures across much of the Pacific are unusually cool, are sometimes blamed for droughts in the Midwest.

When It Rains Too Much . . .

Weather's big winds, the tornadoes and hurricanes, get more media attention because of their made-for-television drama, but the real killers are floods (see Figure 10-6). Even in a hurricane, many more people are likely to perish in the flash floods of its coastal storm surge or its torrential inland rains than are likely to be casualties of its winds.

The widespread flooding that comes when big rivers break out of their banks usually takes an unfortunate combination of weather events, but it only takes a single thunderstorm to make a *flash flood,* a sudden small flood.

Some of the worst floods occur in summer in the United States, especially in the eastern two-thirds of the nation, although flooding is a threat all year long to one part of the country or another.

National Center for Atmospheric Research/University Corporation for Atmospheric Research/National Science Foundation.

Figure 10-6:
Flash flood
in the
Midwest.

Great storms of winter can overwhelm the big river systems when the land they drain won't absorb any more water and they are already carrying the loads of run-off from earlier rains. This happens sometimes when the east-ward progress of big frontal systems becomes stalled, and they concentrate their rainfall in one place.

In the west, rivers and streams flood occasionally when unusually heavy rain-storms follow warm temperatures in late spring. The rain falls on melting mountain snow, filling streams and rivers with more water than their banks can contain. In mountainous terrain, sudden torrential downpours can quickly fill small streams in narrow canyons with dangerously charging water.

East of the Rockies, flash floods are a common threat from severe thunder-storms in spring and summer. Flash floods and more widespread river flood-ing can occur especially when the severe thunderstorms organize themselves into the big systems described in Chapter 9.

In the Southwest, floodwaters can sweep down into the *arroyos,* the normally dry riverbeds of the desert, ripping across highways and into housing subdi-visions that have been built in harm's way.

The felony summer of '93

If bad weather were a criminal offense, the summer of 1993 would be in a dark cell of a bad prison. The season made a terrible mess of most of the eastern two-thirds of the United States. The Midwest was devastated with flooding of historic proportions, and the Southeast was hit with heatwaves and drought.

Weather scientists suspect that El Niño, warm water in the tropical Pacific Ocean, was an accomplice, but they can't prove it. In any case, all summer long, low pressure hung high over the Great Plains, along with an especially powerful jet stream, and a giant Bermuda High pressure system blanketed the Midwest with warm moisture. Hoards of supercell thunderstorms dropped softball-sized hail in places, spawned nearly 650 tornadoes, twice the normal number, and dumped torrential rains.

Widespread flooding along hundreds of miles of the Mississippi River system tore out bridges, swamped millions of acres of farmland and several towns (see figure). Caught in the dry eye of the Bermuda High, the Southeast had the second driest July on record and some of the hottest temperatures ever.

At the end of the summer, at least 100 people were dead from the effects of the heat in the Southeast and Northeast. Another 48 were dead from the floods. And the bill for the losses to property and crops for the season came to more than $16 billion.

National Center for Atmospheric Research/University Corporation for Atmospheric Research/National Science Foundation.

The upper Midwest faces a special threat when melting ice jams form temporary dams that unexpectedly force flows from heavy spring rainstorms over riverbanks.

All across the country, spreading real estate development raises the risks of flooding. Overflows that used to be harmless flooding of rural fields can become major threats in areas of urban sprawl. Streets and highways and parking lots, houses and buildings take the place of ground that used to absorb the rains, sharply increasing runoff from storms. Experts estimate that real estate development increases rainfall runoff two to six times over what would occur on natural terrain. Housing and other development built in natural floodways pinches off the normal flows of small streams at the same time it raises the amount of storm drainage flowing into them. Urban storm drain systems can be quickly overwhelmed by the sudden downpours of severe thunderstorms, turning streets into swiftly flowing rivers.

The tropical storms and hurricanes of summer and autumn, which Chapter 7 describes, pose special flooding threats to coastal regions around the Gulf of Mexico and along the eastern seaboard. Tidal surges combine with heavy rains from these giant storms that can choke coastal rivers and bays. Escape routes can be cut off by high water. Flooding threatens coastal communities and inland areas long after hurricanes lose their most powerful punch as they come ashore. In fact, flooding is a much bigger and more common threat from a hurricane than its powerful winds.

Widespread river flooding causes more property damage, but flash floods, the sudden rising of small streams and the swamping of streets, are more lethal to humans. They can develop in a matter of seconds. Floods kill about 100 people a year in the United States. Many victims are caught in their sleep when floods strike suddenly at night. Many others, nearly half of all flash flood victims, are caught in their cars and swept into floodwaters when they try to cross flooded areas of streets and highways.

When It Rains Too Little . . .

The long periods of below-normal precipitation known as *droughts* are common circumstances that are different from all other weather events. Droughts are the result of something that is expected, but *doesn't happen*. Like floods, droughts occur almost every year somewhere in the United States, and their seasons depend on where you live. No matter where you live, however, summer is the season when the effects of drought always are most noticeable. This is the growing season when high temperatures evaporate the most moisture from the soil and everybody uses more water.

Wildfire season

Wildfire seasons are underway somewhere in the United States just about every month of the year. Because winter is the dry season in Florida, wildfires are a hazard very early in the year. Early spring is often the dry wildfire season over much of the East Coast, including New England, while the Midwest has wildfire seasons in the spring and fall. The South is often most at risk from wildfires in the fall, while the western states have a long wildfire season through the summer and autumn. Santa Ana winds, which Chapter 4 describes, can fan wildfires in Southern California clear through December.

Wildfires are affected by the kind of long-term seasonal differences you think of as climate factors. An especially dry or wet winter or spring or summer, for example, raises or lowers the risk of fire, and wildfires are more likely in drought conditions anytime, everywhere. Studying climate conditions, forecasters often can predict well in advance what level of seasonal fire danger a region might expect.

Wildfires also are affected by immediate weather events. In a matter of minutes, a rash of lightning storms can set off wild fires.

Bad fire weather conditions can combine with human error for disastrous results. In May 2000, a "controlled burn" fire set by the National Park Service to clear brush at the Bandelier National Monument flared out of control in New Mexico and burned for days. More than 400 famiilies saw their houses destroyed in and around Los Alamos. Park Service managers had failed to account for high winds forecast for the region.

During fire season, a National Weather Service forecaster is assigned to predict every day how weather conditions are going to affect the danger of wildfires in a region and also how such conditions and winds and humidity will affect the ability of firefighters to battle such fires.

Large wildfires can create their own weather. Their hot updrafts can cause cumulus clouds to form, even causing thunderstorms that occasionally bring rain and hail. The powerful updrafts can change the circulation of the winds, causing air to rush in toward the fire and fanning it into an explosive firestorm.

In some areas of the West, where summers are normally dry, a coming drought can be identified months in advance. Where the water supplies of farms and large urban populations depend on water from melted snow that is stored behind big reservoirs, you can see drought coming as soon as the winter snow season ends. When the depth of the mountain snow pack is measured and found to be too small, the word goes out to all water users who are downstream of the reservoirs. Farmers are warned that they will not receive their normal supplies for crop irrigation, and people in cities are warned to expect rationing or other limitations on uses of water that are not considered necessary when times are dry. Southern Californians may suddenly realize that they live in a desert. Swimming pools and backyard spas may go empty, and lawns may go brown.

In the eastern two-thirds of the United States, where summers are normally wet, drought most often first shows itself in the heat of the season. In the regions that get much of their rainfall and water supplies in summer, drought is something that creeps up on you with the passing of the long days of dry skies. When you first hear the D-word, usually it's from a farmer.

Farmers who keep their eyes on the moisture of the soil are usually the first to notice the signs of drought and often are its biggest victims. City slickers may hardly notice a drought that can wreck a farmer's year. Even a dry spell of several weeks that may not be generally considered a drought can ruin a wheat crop in the Great Plains or a corn crop in the Midwest if it comes during the summer growing season. An *agricultural drought* is generally thought of as a lack of sufficient water for a particular crop, while a *meteorological drought* is a longer term condition of below-normal rain.

The Dust Bowl

Droughts have come and gone many times over the United States, but one stands out more than any other in history and literature for the deep scars it left. It is a series of four droughts that combined over the southern and central Great Plains and lasted the whole decade of the 1930s. It came to be known as the Dust Bowl (see figure).

Farmers plowed and planted their fields, but nothing would grow. Winds picked up the soil in great clouds that darkened the sky into "black blizzards," and it rained nothing but dust over the region for months on end. Desperate families were forced from their land, and thousands crossed the desert into Southern California in search of food, a job, and a new start. The famous novel by John Steinbeck, *The Grapes of Wrath,* is a story about such a family moving west to escape the Dust Bowl. The Dust Bowl experience changed the country. It changed the way farmers treat the land in the Great Plains, from Texas to North Dakota, where the soil is subject to wind erosion. The plight of the farm families changed the way the government treated the poor, inspiring several "New Deal" relief programs of the era of the Great Depression. And it changed the way weather scientists think about droughts — how long they can last, and how bad they can be.

National Oceanic and Atmospheric Administration/Department of Commerce.

East of the Rockies, droughts and floods are related to one another in a way. It's a feast-or-famine kind of thing. The biggest threats of flooding are from the giant complexes of thunderstorms that get organized over the region, where 50 or more of them might be expected through the summer months. And yet, the farmers of the Great Plains and the Midwest rely on these storm complexes for 80 percent of their growing season rainfall.

In poorer nations, drought can lead to food shortages or *famines* that cause widespread disease and starvation. In the past, as Chapter 16 describes, millions of people have died in the famines that followed droughts in India and Africa. Hardly a year goes by, it seems, when drought does not threaten lives in some parts of the world.

In the United States, droughts affect prices for some foods as suppliers purchase goods that are in short supply on the international market. The economic losses caused by drought in the United States are huge. Experts peg the yearly losses at $6 billion to $8 billion — on average, far more than any other weather events. Floods cost $2.4 billion a year, on average, and hurricanes between $1.2 billion and $4.8 billion.

Chapter 11

Falling for Autumn

• •

• •

Spring gets all of the credit for making the world go around, all of the good press coverage about the seasons. It's that sense of a new year of growth and all. And, of course, you know how the poets are about the birds and the bees that time of year. But really, when it comes to plain good weather, in many parts of the United States autumn is the loveliest time of year, especially the early part of the season.

There's a cool, dry, restful *stability* about early autumn that the other seasons can't match. There's no time like early autumn, after the excesses of summer have left and before winter shows its face. Take October — now there is a month of nice, peaceful weather. No more summer thunderstorms. In fact, October is the driest month of the year from Portland, Maine, all the way along the East Coast and around into the Gulf Coast to New Orleans, Louisiana. Even Florida is drying out.

There are exceptions, of course. That's one of the interesting things about weather. Just when you say something nice about a month, somebody can remember something nasty about it. As Chapter 7 describes, hurricane season runs right on through October, for example, and in fact, the Gulf of Mexico and the Caribbean Sea can be especially dangerous then. And it was in October 1991 when "The Perfect Storm" hit New England. Still, usually autumn is nice, although nobody calls it The Perfect Season.

This chapter describes the variety of weather around the United States during this period of transition between summer and winter. You find out what's behind the famously brilliant fall colors of the leaves in the hardwood forests of New England. You get a look at "Indian Summer" and the fogs of fall and even a peek into the weather behind the Perfect Storm.

The Timing Thing

Right down to the minute, astronomers can tell you when autumn begins — always on or about September 22. The Sun's rays are beaming down smack over the Equator. As far as the Northern Hemisphere is concerned, those rays are heading south. The tilting Earth is at the midway point in its orbit of the Sun between June 21 when the Northern Hemisphere gets its rays at the max, and December 21 when the northern half is tilted away and gets the least of the Sun's warmth. As Chapter 2 describes, it is the day of the *autumnal equinox*, the time, like the first day of spring, when every place on Earth has 12 hours of daylight and 12 hours of night. That's what the Latin word *equinox* means — equal night.

But down in the mess of gases known as the atmosphere, as usual, a lot of other stuff is going on. Big changes are at work in the atmosphere. Cold air is on the move. Before long, as the days grow shorter, the night of the first frost arrives across the northern states from the Rockies to New England. And the leaves on the trees in the hardwood forests catch fire with the brilliant colors of autumn.

Falling Highs and Lows

Dramatic changes are taking place this time of year in the giant high and low pressure systems that influence weather all over the country.

In the Pacific Ocean, centers of air pressure are shifting in a big way with the beginning of fall. The *Pacific High,* which has been an important feature of much of the summer weather across the western U.S., has started shrinking and dropping south. Before this season is out, this high usually will have dwindled to a shadow of its former self and will have migrated to the warmer waters off the coast of San Diego.

The north Pacific's high pressure, meanwhile, is being replaced by a low pressure system known as the *Aleutian Low* that is a big feature of late fall and winter weather over much of the country. The winds flowing in the counterclockwise rotation of this big low pressure system sling many of the Pacific storms of late fall and winter far over the continent.

In the west's Great Basin, high pressure is beginning to replace a large region of low pressure that built up above the summer heat. Around this high builds a new circulation that has far-flung effects — fanning wildfires, turning around ocean breezes, and pushing fog away from the West Coast.

In the lands of the far north, the chilling effect of the fall's dwindling sunshine waken the polar highs over Canada, and these masses of cold, dry air begin to grow and intensify and to slide farther and farther south. Many years, the migration of this air brings welcome relief to regions of the Midwest from the soggy days and nights of summer.

In the Atlantic, the big *Bermuda High* that has kept tropical moisture flowing up over much of the eastern two-thirds of the United States shrinks in size through autumn and migrates to the east. (In fact, it is called the *Azores High* this time of year.) Still, when low pressure troughs developed in the upper atmosphere this time of year, the high pressure in the Atlantic still can bring an unseasonably mild and humid southerly flow to the East.

Coast to Coast

Like all the seasons, autumn has a very different look and feel in different regions across the United States. Early autumn is especially different from east to west.

Most of the East is beginning to dry out from the showers of soggy summer. As a rule, the sky of early autumn is cooler, drier, and more stable than much of the rest of the year. With less heating, summer-type thunderstorms are far less frequent. And early in the season, usually conditions are not yet right for the big mid latitude storms that will bring winter's precipitation.

The biggest weather threats in the eastern and southern parts of the United States this time of year most often are tropical storms and hurricanes, which Chapter 7 describes. Such a storm can turn the region very wet in a hurry. Some years there are several, and some come ashore, and some years they don't. The tropical storm season lasts through October. Riding up the East Coast, once in a while these storms collide with cold Arctic air and generate fierce wintry storms.

Across parts of the West, the early autumn sky may be showing the first serious clouds to be seen in months. The region is saying goodbye to its long dry summer and finally there's a hint of moisture on the way.

In the rainy Pacific Northwest, another wet season can arrive very quickly. The jet stream is beginning to slide back south of the Canadian border and Pacific storms are finding their tracks back down over the region.

Across the northern plains, and from the upper Midwest to New England, cool air comes quickly with the fading light.

In the upper atmosphere, the westerly winds that will bring winter storms are building strength as autumn wears on. Winter storm tracks begin appearing, and before autumn is done, winds will howl across the Great Lakes, and snow will fall over much of the Midwest and Northeast.

By the end of autumn, the winter storm tracks are well established, the westerly winds and the jet streams are flowing far south over the United States, and most of the country looks and feels a lot like winter.

In a Pigment's Eye

A funny thing about the famous fall colors: It's not so much that the leaves *turn* yellow or orange or purple or red as it is that they *stop turning green*. The pigments that bring those brilliant autumn hues to the trees that drop their leaves were there under the green. When the leaves are cooking food for the plants, they produce a lot of green pigments that hide the other colors. As fading sunlight and cooling temperatures signal the trees and leafy plants that it's time to shutdown food production, those green pigments pack it in for the year.

Not all dying leaves knock your socks off, of course. The drab brown of a shriveling elm is not much to write home about. Maples, on the other hand, make real spectacles of themselves, although they differ from one species to another. Leaves of red maple turn brilliant scarlet. Sugar maples become orange and red. Black maples are a glowing yellow. Hickories are golden bronze. Aspen and yellow-poplar become golden yellow. Dogwood turns purplish red. Leaves of beech are light tan. Sourwood and black tupelo turn crimson. Oaks turn red, brown, or reddish-brown late in the year after most trees have lost their leaves.

Experts have given up the old idea that frost promotes these changes in color. The best seasons of fall colors, especially the red and purple pigments, come from a lot of autumn days that are warm and sunny and autumn nights that are cool and crisp — but not freezing. Soil moisture is another key factor. The stress of droughts makes for dull autumn leaves. A warm, wet spring can lead to an especially brilliant fall.

Indian Summer

The idea of a welcome warm spell that comes fairly often, but not always, after the first killing frost and stretch of cold autumn weather is as old as the hills. Why the early European settlers in the United States gave it the name "Indian summer" is a factoid that seems to have fallen through the cracks of history. They could have called it "St. Luke's summer" or "St. Martin's summer," which are common names for it in England. Elsewhere in Europe, it was "Old Wives summer."

What was so perfect about it?

Big Nor'easters are a dime a dozen off the coast of New England. There was a doozy known as the 1962 Columbus Day Storm, for example, and the March 1993 "Superstorm." So what made the big bruiser that whipped up during the last week of October 1991 "The Perfect Storm"?

Weather forecasters say three ingredients came together at the same time in the North Atlantic to produce the storm that became the subject of the best-selling book and movie.

✔ A large high pressure system over northern Canada supplied a large pool of cold air.

✔ A weak low-pressure system was traveling along the front where this large cold air mass pushed off the coast of New England on October 27.

✔ Flowing up from the Tropics was the fuel that gave this Nor'easter its epic proportions: the warm, moist air of a dying hurricane.

Retired National Weather Service meteorologist Bob Case was on duty at the Boston forecast office in October 1991 when this whopper was taking shape, and here's how he describes the situation:

"When a low pressure system along the front moved into the Maritimes southeast of Nova Scotia, it began to intensify due to the cold dry air introduced from the north. These circumstances alone could have created a strong storm, but then, like throwing gasoline on a fire, a dying Hurricane Grace delivered immeasurable tropical energy to create the perfect storm."

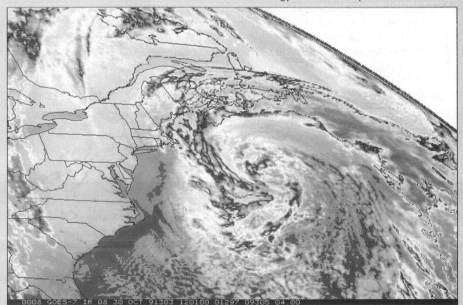

Courtesy of National Oceanic and Atmospheric Administration/National Climatic Data Center.

The more northerly the climate, the more severe the oncoming winters, the more cherished the tradition of Indian summers. It's not just a long summer. The idea is that the temperatures of the coming winter have made themselves felt, and then along comes this last welcome visit of summer warmth at a time of year that could just as easily be stormy and freezing cold. Indian summer skies are clear and its days sunny and often hazy.

By whatever name, Indian summer reflects something that also is true of spring, that other fairly unpredictable season of transition. The reasons for the seasons are the steady, clockwork motions of Earth orbiting the Sun at a tilt. But down under the atmosphere where weather happens, the gains and losses of the Sun's energy come and go in fits and starts, like advances and retreats of opposing armies.

In the Fogs

The stable, cooling air of the season and the evaporation of summer's moisture from the ground can make autumn a foggy and treacherous time of year over the Northeast and much of the Midwest. Because the water now is warmer than the air, steam fogs that Chapter 3 describes often form easily over many lakes great and small. Dewpoints still are much higher than they will be when winter sets in, so the air condenses into fog more easily than it will when colder temperatures arrive.

After the fog forms during the long, cool night, the next day's sunlight may be too feeble to evaporate it very quickly. The result can be that children are waiting for school buses in the dark early morning fogs of October before clocks are switched from daylight saving time. Many school systems in the region call for two-hour delays in the start of school on such mornings.

Foggy autumn mornings make all travel difficult and highway travel often dangerous. Some of the worst chain-reaction traffic accidents occur in these conditions. Air travel can be seriously disrupted. Fog can cause delays on some flights that have a ripple effect throughout the nation's air traffic system.

The same conditions that bring fog can lead to heavy dew across the countryside. As Chapter 3 describes, dew is formed when water vapor collects on the ground, blades of grass, and other surfaces as it comes into contact with cold night air.

Fires of the Wild West

Autumn can be a dangerous time in the West. Most of the countryside is hot, and it hasn't seen rainfall in months. It is fire weather, and forecasters across the region always find themselves in the thick of it.

As Chapter 4 describes, the strengthening off-shore flow creates dry, hot Santa Ana winds that can fan wildfires through the mountain canyons of central and southern California.

The desert brushlands are especially dry this time of year, and when the Santa Anas are blowing down from the high desert — downslope winds that are warming as they fall — the slightest spark can grow to a raging firestorm in a matter of minutes.

Weather forecasters carefully follow wind and other weather conditions this time of year in the West to help firefighters watch for hazardous fire conditions. The fire hazard can extend far into the season before rains finally quell the danger in southern California and the Southwest.

Part IV
The Special Effects

In this part . . .

*I*t's your weather. Nobody sees it at quite the same angle that you do. When you look at a rainbow, for example, and marvel at its colors, you are the only one seeing that particular rainbow. Even a person standing next to you sees it differently, the light bouncing back from different raindrops. And your friend across town, she may not see it at all.

One of these days you're going to look up and the sky is going to surprise you again with a new effect of light or color. It's always doing that, showing off for you, so keep an eye on it.

In this part is a wide variety of special effects the sky has in store for you. And you can find some cool things you can do at home to make some of your own weather effects and to keep track of changes in the weather around you. By making weather a personal thing, you're following in some big footsteps. In this part, you can find out who they belong to.

Chapter 12

Taking Care of the Air

● ●

In This Chapter

▶ Seeing through the haze

▶ Plugging the ozone hole

▶ Warming to the greenhouse effect

● ●

Would you buy a dirty book? I mean, would you have forked over your hard-earned cash for *Weather For Dummies* if there had been *stuff* on its cover? If it smelled funny? If some of its pages were smeared to the point that you could barely read its wonderful prose? No way — not unless your life depended on it. It's a silly question, in a way, but look what you and I put up with when it comes to something that everybody's life *does* depend on. Imagine that you had to pay for the air that fills your lungs. Would you put up with all of that *stuff?* By the way, it's your book, but *whose air is it?*

This chapter is all about the stuff in the sky that didn't used to be there and doesn't belong there. It's about all forms of air pollution and what the weather has to do with it. This is the stuff that makes that thick layer of air over a city look like a gaggy brown blanket of mud. The stuff that stinks and burns your eyes and makes you cough and short of breath. The stuff that makes the atmosphere do weird things like developing ulcers in its ozone layer and becoming all gassy and maybe even overheated.

It's about soot and smoke. It's about the stuff that comes out of the back ends of motor vehicles and the tops of factories. *Emissions* is the polite and sanitary word for it. My people at the Go Figure Academy of Science are always looking out for words like emissions. You could say, for example, that when you're following a herd of cows, you carefully watch where you are walking so that you don't step in any of their emissions.

Air pollution raises some difficult questions. Is the sky a really good waste dump? Does it all come out in the wash of a good rainstorm? Is it true that, year after year, it can take in millions of tons of fumes — these emissions — and the atmosphere will always find something harmless and convenient to do with them? As the arguments over global warming show, weather scientists don't have all the answers to these questions yet.

Polluting the Air

You can't blame the weather for polluting the air, but weather conditions certainly can make matters worse. Fog and stagnant air produced the lethal London fogs that were mixtures of smoke and fog (see sidebar). This combination of smoke and fog is where the word *smog* comes from. Heat and sunlight cook up a different brew called *photochemical smog* that car-crazy Los Angeles made famous in the 1950s.

Most often, episodes of heavy air pollution are created when air temperatures are in a pattern known as a *temperature inversion*. It gets this name because the air temperatures from the surface to upper layers are upside down, or inverted, from their most common pattern. Instead of growing cooler and cooler with higher and higher levels, there is an *inversion layer* of air that is warmer than the air nearer the ground. When air is warmer at the surface, it can rise through the cooler upper air like a balloon, dispersing the pollution it carries and creating some circulation. During a temperature inversion, the layer of warmer air is like a lid, preventing the cooler air near the surface from rising and carrying away its polluting gases.

Temperature inversions are common any time there is a pattern of high pressure and sinking, warming air aloft. You could think of inverted air temperatures as a salad dressing of vinegar and oil. They sit there in separate layers unless a wind or a storm comes along to shake them up.

A temperature inversion is very common along the West Coast of the United States, where the worst episodes of air pollution are during summer. The prevailing wind blows in air from the cold Pacific, and the cool air often gets trapped under a layer of warmer air. Cool air also can slide down the western mountains and sink under warmer air over the valleys.

Elsewhere in the United States, including the Rocky Mountains, the pollution season often is winter, when cold surface air often gets trapped below warmer air overhead. In these parts of the country, the worst episodes of air pollution often come with a layer of fog.

Stuff that pollutes the air comes in several varieties. Many of them are gases. Several are what are called *greenhouse gases,* which trap heat like the roof of a greenhouse and prevent it from radiating out into space. These gases are what *global warming* is all about — concern by many of the world's leading climate scientists that the greenhouse gases are building up to the point where they could be causing the atmosphere to warm up more than it would naturally.

The short and long of it

Air pollution problems come in different varieties, too. The same puffs of pollution by the same gases have more than one impact. It's the time frames that are different.

In the short run, people in many large cities face immediate health hazards in the air they breathe.

- Day-to-day weather conditions like temperatures and cloud cover gang up with the stuff drifting around in the local air to pose one kind of public health situation or another.

- From one season to another, air pollution comes and goes with the winds, the passage of storms, occasions of high pressure, hot temperatures, or fog.

In the long run, the same polluting gases also have effects on health and climate, the Big Picture averages of weather.

- Some polluting gases continue to break down the ultraviolet shield protecting Earth for decades or even centuries after they have been dumped into the air.

- The long-term future also is clouded by concern over the added greenhouse effect of several air polluting gases on climate patterns around the world.

London smog

Smog was invented in London, you might say, along with the Industrial Revolution.

People had been complaining for hundreds of years about the smoke problems in England, especially since the 1600s, when coal replaced the shrinking wood supply as fuel for heating homes. As the smokestacks of new industry sprouted over London, the sky grew darker and darker. It was the lethal brew of London's smoke and fog that inspired a physician there to coin the word smog.

It was 1911, and 1,150 Londoners had just died of an episode of air pollution. The "smoke problem" was considered just part of modern life in those days, the price of progress and profits. For another 50 years, nothing much was done about the problem.

In the first week of December 1952, high pressure built up overhead, and the winds died. Another bout of stubborn London fog set in. The city grew dark, and the air grew dense. Visibility was down to 1 foot. People wore masks over their faces and made their way down the street by feeling the sides of buildings. By the time the fog lifted five days later, 4,000 people were dead.

Parliament approved Britain's Clean Air Act in 1956.

The usual suspects

Sorry to get technical about this, but here are some names of air pollution suspects that you may want to know:

- *Carbon dioxide* is a big part of the pollution problem, even though it occurs naturally in large quantities in the atmosphere. It's the stuff you breathe out before you take another breath of oxygen. Everything that burns pumps more CO_2 into the sky. A lot of carbon dioxide gets washed out of the atmosphere when it rains or snows, but not all of it. For 150 years or so, ever since factories and vehicles have been burning coal, oil, and natural gas, this greenhouse gas has been building up in the atmosphere.

- *Hydrocarbons* are greenhouse gases that are combinations of hydrogen and carbon that spew out of the tailpipes of cars as well as many oil refineries and factory smokestacks. Many of them are bad for you. Some of them cook up with oxides of nitrogen in the air on a hot day to form that brown layer of smog.

- *Nitrogen oxide* is another tailpipe emission that forms the brown haze of smog and acid rain that eats away stuff it falls on. It's bad for your heart, lungs, liver, and kidneys, and it helps cause pneumonia and bronchitis.

- *Ozone* is an unfortunate byproduct of the processes that brew air pollution. This stuff that you and I like so much up in the stratosphere, where it forms the protective layer against harmful ultraviolet radiation, is definitely not something you want in the air you breathe. It irritates your eyes and lungs and is toxic.

- *Sulfur compounds* come from burning coal and oil, especially by electrical power plants, and from paper and pulp processing plants. They make the air stink and cause acid rain. They are bad for everybody, especially people with breathing problems like asthma and bronchitis.

- *Suspended particulates,* or floating particles, are tiny bits of stuff that includes metals like lead and zinc and copper, as well as really small bits of soot and stuff like asbestos fibers and bits of fertilizer and pesticide that are so small they can be inhaled deeply into your lungs. Also in this floating mix are *aerosols,* tiny liquid droplets of acid and other substances that you wouldn't want in your soup.

Getting the drift

Winds play a big role in air pollution. Strong winds help blow away and dilute air pollution where it is created. In fact, many areas where a lot of pollution is pumped into the sky very much depend on prevailing winds to keep this stuff on the move. Many of the worst cases of air pollution have come when the prevailing winds stopped blowing.

Donora 1948

On October 26, 1948, a big area of high pressure settled over the eastern United States. Right in the middle of it, in the bottomlands of the Monongahela River Valley, was the little town of Donora, Pennsylvania. Smokestacks of the town's steel mill and zinc smelter and sulfuric acid plant were pumping out their stuff just like they always did. The air was moist from recent rains, and a strong temperature inversion put a lid of unmoving air over Donora.

For five days the foul air sat over the town, and all the time the smokestacks kept pumping out their stuff. A thick layer of dense fog blocked out the light of the Sun. It was so dark, you could barely see across the street, and the whole place stank.

Before it was over, 22 people were dead, and half of the town's 14,000 residents were sick. Finally, an approaching storm came along and blew it all away.

Official reports of the worst air pollution disaster in U.S. history blamed the weather. But what happened in Donora in 1948 woke everybody up to the dangers of industrial air pollution and eventually led to enactment of the federal Clean Air Act in 1963. Today, a marker in the little town pays tribute to the air pollution victims of Donora.

During the big 1997–'98 El Niño, for example, South Asia was shrouded in a cloud of smoke from fires set to clear the jungles of Indonesia. The farmers had expected the prevailing westerly winds to blow away the evidence of their fires. But El Niño calmed the winds, and for weeks the smoke built up to produce a widespread health crisis in the major cities of the region.

Winds play another role, and scientists are discovering some interesting things about them. More often than not, prevailing winds deliver the effects of pollution to areas downwind of the polluter. But there's more to it than that. Storms can blow air pollution very high into the atmosphere, where the jet stream and other powerful winds can carry it great distances around the world.

The sky over places like the Grand Canyon is showing the telltale signs of pollution from air blown in from distant cities. During the 1997–'98 El Niño, smoke from wildfires burning in severe drought conditions in central Mexico drifted as far north as Chicago. Recent studies show that the sky over the Pacific Northwest from time to time contains traces of stuff pumped into the air in Asia.

Acid rain

Don't you hate it when it rains acid? Figure 12-1 shows the concentration of acid rain in the eastern third of the United States as well as southeastern Canada. Rain is normally a tad acidic, because water droplets in clouds dissolve some of the naturally occurring carbon dioxide in the air. *Acid rain* refers to concentrations of acid that you'd expect to find inside of a tomato. (Weather tip: When it rains tomato juice, go inside.)

In thousands of lakes in the upper Midwest and eastern Canada — a region that is famous for its good sport fishing and its natural beauty — acid rain has killed fish populations and increased algae blooms. In more than 1,300 streams in the Appalachian Mountains, the water is acidic, and fish populations have declined. From Georgia to Maine and across eastern Canada, it has caused forests to decline. In cities across the region, it has been eating away at the sides of buildings and gnawing through paint and dissolving metals on things like automobiles.

Unlike a lot of things, most everybody knows what the problem is with acid rain and where it comes from. The biggest problem is emissions of sulfur compounds that produce sulfur dioxide that rains out of the sky as sulfuric acid. A recent study found that coal-fired power plants of electrical utilities spew 70 percent of the sulfur dioxide that's in the air. One study found that such plants in the Ohio Valley are responsible for half of the acid rainfall in Canada.

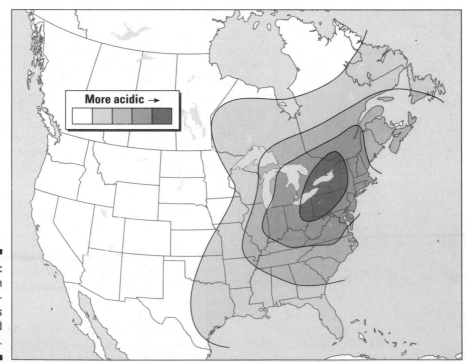

Figure 12-1:
The pattern of acid rainfall across the U.S. and Canada.

Nitrogen oxides also make acid rain. About a third of these emissions comes from power plants of electric utilities, and another third comes from motor vehicles.

Over 20 million tons of this stuff — sulfur and nitrogen compounds — is dumped into the atmosphere every year. U.S. government programs are underway to reduce these emissions, especially sulfur dioxide, but progress is slow and expensive. Canadians don't seem to be very impressed with the pace of things. The fish aren't talking.

The Hole in the Sky

Sunlight is pretty powerful stuff. After all, it's radiation coming from the nuclear fires of a star. If all the different wavelengths of light that left the surface of the Sun were to arrive at the surface of your skin, you'd be deader than a doornail. So would I, of course. In fact, if the atmosphere did not filter out some of the radiation that reaches Earth, most forms of life would not exist.

The killer rays of sunlight are known as *ultraviolet radiation,* because the lengths of their waves are in the invisible range out beyond the color violet that you see. Some of this radiation — called *UV-B* — causes sunburn, skin cancer, and eye damage. (Check out the UV Index in Chapter 10.) The only thing between you and these deadly rays is a thin layer of gas high in the sky above where the weather takes place. It's between 9 and 30 miles up, in the band of sky known as the stratosphere. The gas is a rare form of oxygen known as *ozone.*

You may wonder: Why is a layer of gas that could be more than 20 miles thick in the atmosphere called a *thin layer?* It's because ozone molecules are rare even in the band of sky they call the *ozone layer.* There is very little of this stuff. For every ten million molecules of air in the atmosphere, there are only three molecules of ozone. If you could take the total amount of ozone and spread it evenly around the atmosphere, like icing on a cake, the layer of ozone protecting you and me from the killer rays of space would be only about one-eighth of an inch thick.

The bad news is, because of air pollution, this thin shield of ozone is getting even thinner. Some chemical reactions are taking place up there in the stratosphere that are eating the ozone. It has reached the point to where an *ozone hole* forms in the sky over some parts of the world.

Fortunately, it has been happening over unpopulated Antarctica, the South Pole, during the Southern Hemisphere's spring, September through November. But that's just the first and the worst of it. The ozone thinning problem is spreading to other parts of the world. Now it's showing up over the Arctic, the North Pole, during the Northern Hemisphere's spring, and even over more populated regions.

The chemistry of what scientists call *ozone depletion* involves some fairly complicated chain reactions. It involves human-produced compounds that contain a variety of combinations of these chemical elements: chlorine, fluorine, bromine, carbon, and hydrogen. The ozone-eating process also requires especially cold temperatures and a convenient surface for the chemical reactions to take place, like the side of an ice crystal in a stratospheric cloud.

The end result of all of the complicated chemistry is some fairly simple math. The only equation you need to keep in mind is this: Less equals more. Less ozone in the stratosphere over your head means more lethal ultraviolet radiation in your face.

The ozone-eaters

The chemicals that eat ozone had enjoyed a good reputation during their working years, before people got a look at how they behaved in retirement. Since the 1930s, the gases that contain chlorine and fluorine had been important products called CFCs (for chlorine, fluorine, and carbon) that had a wide range of uses:

- ✔ They were used in refrigerators and air conditioners.
- ✔ They were used in manufacturing blown plastic foam that was used in containers at fast-food joints.
- ✔ They were used in the cleaning of electronics parts and as industrial solvents.
- ✔ They were the favorite gases used to build up the pressure that blew stuff out of spray cans.
- ✔ They were used as the main ingredients in fire extinguishers.

In the 1970s, however, scientists discovered what CFCs were up to long after they had served their intended purposes. Years later, CFCs were mixing in the atmosphere and eventually wandering up into the stratosphere and eating away at the ozone layer. But their warnings about the loss of ozone and what it meant to human health were not taken very seriously for several years.

In the 1980s, scientists discovered the ozone hole over Antarctica. A lot of people thought it was funny and made jokes about the ozone hole. Every year, the problem grew worse. The hole in the sky grew bigger. And then scientists detected something that made the ozone hole jokes about as funny as a case of skin cancer.

In the 1990s, satellite instruments and measurements from high-flying airplanes found the problem had spread to the stratosphere over the heavily populated middle latitudes, where you and I live, and over the Arctic. They found more ozone-destroying chemicals up there and signs that the ozone layer was becoming thinner.

Smog ozone

Ozone is like a big, obnoxious brute who happens to be on your side in a fight. You know, the friend of a friend. You're glad he's there, at a safe distance, but you really don't want to spend a lot of time around him.

While most of the ozone floats around in the stratosphere, where it does a lot of good absorbing ultraviolet rays, about 10 percent of it lays around in the lower part of the sky known as the *troposphere,* where it gets in your face.

This is just about the last stuff you want in the air you breathe, but there it is, the main ingredient of what is called photochemical smog. This is the dry smog of a hot summer day in a city that cooks up a thick brown haze in the middle of the afternoon.

It smells bad, for one thing, and it burns your eyes and makes breathing hard, especially for people with asthma and other conditions, and causes chest pains, nausea, coughing, and even heart trouble. It damages crops and forests.

This smog ozone forms when other gases that come out of the back ends of automobiles react with sunlight.

All these discoveries led to a series of international agreements, beginning in Montreal, Canada, in 1987, calling for reductions in the production, and finally the elimination, of CFCs. More ozone-friendly substitutes have been found for the jobs done by CFCs.

On the mend?

So, if CFCs are no longer being pumped into the sky, is the hole in the ultraviolet shield repaired? Shouldn't you and I be hearing the good news any minute now that the ozone hole is closed? Isn't the threat of more cases of skin cancer and other serious health problems from unfiltered UV-B radiation a thing of the past?

It sure seems reasonable to think so, but air pollution turns out to be a tricky and risky business. Even when everybody responds to the best science around, things that go wrong in the atmosphere can *stay wrong* for a long time. Sometimes it's called "the weather machine," but any forecaster can tell you that the atmosphere doesn't really act anything like a machine. There are no switches to pull or buttons to push up there. The answers to all these questions is *no.*

Nobel Prize science

The ozone hole story is pretty remarkable when you think about it. Imagine trying to get a handle on complicated chemical processes taking place among invisible gases in super-cold temperatures on the surface of icy cloud crystals 10 miles in the sky. Even before the ozone hole began appearing in the 1980s, scientists had figured out that something was going wrong up there.

Their science was so good that by the time the hole was discovered over the South Pole, people knew what the problem was, and what to do about it. The threat was so real, and the idea of losing the ultraviolet shield was so alarming, that people around the world who couldn't agree on much of anything agreed about the ozone hole.

The 1995 Nobel Prize in Chemistry was awarded to three scientists whose pioneering work in the 1970s led the way. They are Sherwood Rowland and Mario Molina, who did their work at the University of California at Irvine, and Dutch scientist Paul Crutzen of the Max Planck Institute for Chemistry in Germany.

Awarding the prize, the Royal Swedish Academy said their work helped avoid a catastrophe.

While scientists can detect the decline of new CFCs entering the lower atmosphere, the CFCs that already have been released into the air can float around in the upper atmosphere anywhere from 50 years to hundreds of years. Leading scientists studying the problem for the World Meteorological Organization estimate that the ozone-thinning process will "gradually disappear by about the middle of the 21st century" as natural processes slowly remove the CFCs.

So it looks like the ozone hole and the thinning process is going to be around for a long time, maybe even longer than 50 years. The latest news from the North Pole is not very encouraging for the regions of the world where most people live — the middle latitudes of the Northern Hemisphere.

A large new international study reports that the atmosphere over the Arctic is getting more and more like the high sky over Antarctica. This is not good news. The especially cold temperatures over the South Pole — 120 degrees below zero — that make conditions right for an ozone hole now are showing up more often over the North Pole. They are finding more polar stratospheric clouds, where ozone destruction takes place, over the Arctic than they expected.

Part of the problem may be global warming, scientists say. While the greenhouse effect raises temperatures in the lower part of the atmosphere, it actually makes things colder in the stratosphere.

The Big Warm

Climate changes all the time. This is a simple but big idea that takes some getting used to for some people — like me — who grew up thinking that the world's long-term weather patterns are permanent parts of life. Sure, the world had gone through its ice ages, its big chills, but that was during its early years. It had gotten all of that out of its system, like growing pains. You know, in the back of my mind I must have thought that the planet had just been figuring out how to make its atmosphere *perfect* for the likes of you and me.

Maybe I wasn't paying such good attention. Anyway, research into the deep past has made the changeable nature of the climate pretty clear. By looking through the layers of ice in cores drilled in ancient glaciers and into the trunks of very old trees and other clever means, researchers have shown that climate has been changing all along. Not only does climate change, but sometimes it changes more quickly than you might think. Instead of centuries of gradual pattern shifts, big changes can come just in a matter of decades. And as far as you and I are concerned, it just as easily changes for the worse as for the better.

And now along comes the big warm, a picture of a future climate that doesn't look anything like an atmosphere perfecting itself for the likes of you and me. It's not just that temperatures become a little warmer. Storms can become bigger or more numerous, some scientists say, because more water vapor is in the atmosphere and more heat energy is moving around. The greenhouse roof in the atmosphere is becoming thicker with heat-absorbing gases, and temperatures around the world are on the rise. Like smog in your face and like the ozone hole overhead, global warming may be another wrinkle in the way air pollution affects the world.

On the natural

Changes in climate are natural. Not all of them, of course. The ozone hole does not appear to be part of the natural ups and downs of climate. But researchers over the years have built up an impressive record of the very distant past that shows that, on the natural, the world's climate has been both warmer and colder than it is today. One of the most difficult problems for climate scientists is deciding which shifts in today's climate are natural and which ones are more likely the result of air pollution by human activities.

Some changes in the world's average temperatures are caused by the way the solar system is put together. Over tens of thousands and hundreds of thousands of years, the Earth's tilt changes and so does its orbit around the Sun. These are the kinds of things that cause ice ages to come and go.

Even quick and dramatic climate change can be natural. When a meteorite or a comet crashes to Earth, if it's big enough the explosion can throw a huge amount of stuff high up into the sky. This dark layer can hang around for years, blocking out a lot of sunlight and changing patterns of weather.

A comet or meteorite crashed to Earth 65 million years ago and led to the disappearance of the dinosaurs, but what exactly killed them? Probably it was the drastic change in the climate. They may have starved as the dark cloud shrank their food supplies, and they may have frozen in the long years of bad winters that followed the great explosion.

Eruptions of very large volcanoes have the same effect. As Figures 12-2 and 12-3 illustrate, for example, the eruption in 1991 of Mt. Pinatubo in the Philippines caused temperatures to fall around the world. The explosions send up great clouds carrying millions of tons of stuff that act like a big sunshade in the sky for several months. Some meteorologists think the eruption was partly responsible for the heavy snow in the eastern U.S. in the winters of 1992-'93 and 1993-'94.

Figure 12-2:
Mt. Pinatubo erupts in the Philippines.

Courtesy of the United States Geological Survey, Photo by J.N. Marso.

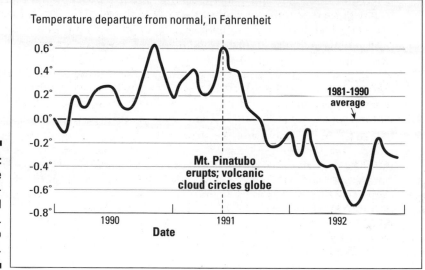

Temperature departure from normal, in Fahrenheit

1981-1990 average

Mt. Pinatubo erupts; volcanic cloud circles globe

Date

Figure 12-3: Average temperatures dipped after Mt. Pinatubo erupted.

Other natural climate changes lasting several months to a few years come from the shifts in tropical Pacific Ocean water temperatures known as El Niño and La Niña, which Chapter 6 describes. Another pattern in the North Pacific shifts temperatures back and forth every few decades or so. These things affect temperatures, the shape and speed of jet streams, the amount of storminess, and the number of hurricanes from one year to the next. Climate scientists are not sure what causes these ocean patterns, but they can tell that they're natural, that they've been around a long time.

On the unnatural

In a way, global warming is a case of too much of a good thing. Just as climate shifts are natural, so is the "greenhouse effect" that people are talking about when they describe global warming. If it were not for the greenhouse effect — the fact that gases in the atmosphere trap heat radiating up from the surface — Earth would be too cold for living things. Water vapor and carbon dioxide are the main greenhouse gases that occur naturally in the atmosphere.

Here's what scientists mean by global warming: It's the heating up of the atmosphere on top of the natural greenhouse effect, which is caused by the gases that humans have added to the atmosphere since the Industrial Revolution began a couple hundred years ago. It's the unnatural rise — and unnaturally rapid rise — in temperatures caused by the fact that you and I have changed the chemical makeup of the atmosphere.

Not much about this global warming business is simple, except maybe its cause: More than anything, it's the gases that are byproducts of the burning of what scientists call *fossil fuels* — coal and oil and natural gas that come from the decay of old plants and animals. They are carbon dioxide, nitrous oxide, and methane. All these are naturally occurring ingredients of the atmosphere, but human activities have raised their proportion and changed the mix.

Figure 12-4 shows you the increase in carbon dioxide in the atmosphere since before the Industrial Revolution. Concentrations of this gas has been measured directly since 1950. Earlier readings come from measurements of air bubbles trapped in ice cores drilled in ancient glaciers in Greenland and Antarctica. Notice the recent rapid rise in the concentration of this greenhouse gas that is caused by human's burning fossil fuels as well as other practices such as burning forests to clear land for farming.

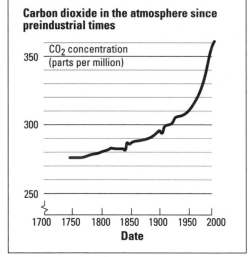

Figure 12-4: The trend in carbon dioxide in the atmosphere since 1750.

The global warming picture

The average temperature of the Earth has warmed 1 degree in the last century, and scientists expect another 2 degrees to 6 degrees of global warming in the next century. That may not sound like much, but it's a lot, and it's faster than anything in the last 10,000 years. Consider this: In the middle of the last ice age, 20,000 years ago, world temperatures were 9 degrees cooler than today. The Great Lakes and New York City were buried under massive ice sheets at the time.

High temperatures mean melting glaciers and ice caps and higher sea levels, for one thing. Already, the oceans have risen up to 10 inches in the last century, and scientists predict a rise of one-half foot to 3 feet over the next 100 years.

The effects of global warming would be different from place to place. Some areas would become drier, and others would become wetter. Droughts and heatwaves would become more common, and air pollution could become worse. At the same time, storms could become more fierce and more frequent in some places. Already, worldwide precipitation over land has increased about 1 percent in the last century, scientists say, and extreme rainfall has become more frequent in the United States.

Scientists can't tell you exactly what to expect, but here's the general picture. Everybody's health, their food supplies, their fresh water, the forests, wildlife, and the coastal communities and wetlands — all these things depend on long-term weather patterns remaining fairly stable, or at least changing only gradually. What worries climate scientists is that all of these things are in harm's way in a world of big, rapid change that is the picture of global warming. Figure 12-5 is a look at the global temperature picture in modern times.

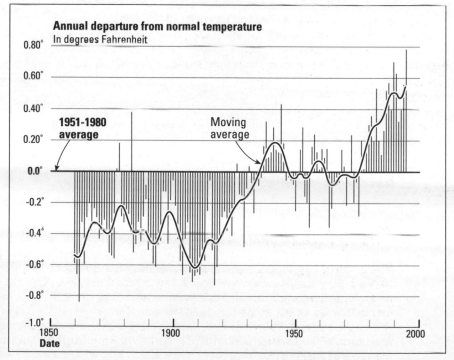

Figure 12-5:
Average world temperatures since 1861.

Annual departure from normal temperature
In degrees Fahrenheit

The global warming debate

Media people who kick around the phrase "global warming debate" are not really describing how the world's leading climate researchers spend their time on this subject. They know that global temperatures have been on the rise for many years now, and most of them agree that the atmosphere is showing signs of humankind's tinkering with its basic chemistry by polluting the air. Their models and their research findings mostly point in the same general direction, but they disagree about the details. There is general agreement among these climate researchers, but still a lot of uncertainties. That's the way it is with science.

Climate scientists argue about things like this:

- Is *this* wrinkle of climate change part of natural variation or human-caused global warming? And how about *that* one?
- Will global warming mean more high clouds that reflect sunlight and cool off the atmosphere, or will it mean more low clouds that absorb more heat from the ground?
- My model is better than yours.

Politicians — now those are the folks debating global warming. Depending on their persuasions and the special interests they represent, some are inclined to fear the worst, and others are inclined to deny it altogether. And both sides are absolutely certain they're right. That's the way it is with politics.

Politicians argue about things like this:

- Reducing these polluting gas emissions is going to cost a bundle, so whose going to pay for it?
- So whose air is it, anyway?
- Oh, yeah? Well, so's your old man!

The heat is on

Is the unfriendly face of rapid global warming showing itself in the quickening pace of rising temperatures around the world? As the decade of the 1990s came to a close, climate scientists were tempted to think that warming may be shifting into a whole new high-speed gear.

As Figure 12-6 illustrates, the global temperature record shows an average warming of about 1 degree over the last 100 years, which was the warmest century in the last 1,000 years. But look what happened in the last two decades. The ten warmest years of the last 100 occurred since 1983, and seven of them were in the 1990s. The 1990s were the warmest decade, and 1998 was the single warmest year of the past millennium.

These records were set even in a decade that saw the giant cloud of Mt. Pinatubo's volcanic eruption fill the atmosphere with its cooling, sun-blocking dust for two years or more. And 1999 was the fifth warmest year on record even as the cooling influence of La Niña conditions in the tropical Pacific Ocean held sway.

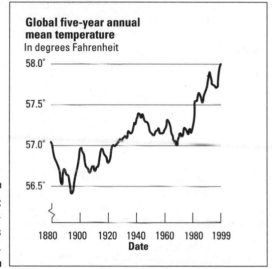

Figure 12-6:
Global temperatures since 1900.

A splash of cold water

Whether or not the air polluting effects of human activities are leading to an unnatural and serious global warming of the atmosphere does not appear to be an issue that is going to be resolved anytime soon. Even if most researchers involved in the subject agree generally among themselves, large numbers of practicing meteorologists, among others, remain skeptical of the whole idea.

Many of them remember, for example, that back in the 1970s, just the opposite idea was in style. After a bout of especially cold winters, scientists and the media began talking about the idea of a coming ice age. There were numerical models, historical studies — the works. There were all kinds of speculations about the coming global cooling.

Sadly, today's global warming debate arouses strongly held views and almost immediately becomes a hot political question rather than a cool scientific one. In the middle of all the heat and emotion, it's easy to be like the people who laughed at Galileo 400 years ago when he argued that air actually has weight. It's hard to keep an open mind.

Wild and crazy

Does the weather seem to be getting crazier and more extreme? Maybe it's my imagination. Maybe it's just a run of bad luck. Or maybe it's just a false impression left by the fact that we are so saturated with media coverage of every event in this day and age. If it's a sign of global warming, when more extreme events like drought, flooding, hail, hurricanes, and tornadoes are expected, my advice is head for the hills and take your wallet with you.

Look how often you and I end up having to pay for the effects of extreme weather. When it rains too much and floods, your tax dollars and my tax dollars pour in to bail everybody out. When it rains too little, you and I pay at the store for food at higher prices. And when a big hurricane comes along, or a terrible winter storm, one way or another, you and I pay for it.

When a hurricane wipes out a coastal community, naturally everybody feels sorry for the victims. But when the national disaster relief program kicks in, questions come up: Why are my tax dollars being spent so that people can rebuild on low-lying coastal land that is in harm's way for the next big storm? The same kind of question is raised after a flood. Why are my tax dollars being spent so that somebody can put another house in the historic floodplain of that river?

The big climate flop

Climate scientists have recently realized that the circulation of the ocean plays a big role in long-term weather patterns. While temperature trends are especially slow to take hold in the ocean, the big deep has a big climate trick up its sleeve.

A great deal of heat energy of the Sun is distributed from the Equator to the poles in a huge conveyor belt of ocean circulation that is controlled by water temperatures and *salinity*, or salt content. Warm water flows north along the surface of the Atlantic Ocean, and then when it cools off, it sinks to the bottom of the ocean and heads south.

Where the water is warm on the surface and where it is cool has a lot to do with the shape and intensity of weather patterns around the world.

The conveyor belt has different modes, or operating speeds, that change the location of rising and sinking ocean water, and rather than slide gradually along, it tends to flop suddenly from one mode to another and stay in the new mode for hundreds or thousands of years.

If globally warming temperatures cause the oceans to heat up, and if melting ice caps change their salinity, there's some chance that a new long-term ocean circulation-climate pattern could kick in. The trick is to figure out how close the conveyor is to a flop.

Chapter 13

Up in the Sky! Look!

● ●

In This Chapter

▶ Bending with the light

▶ Taking apart the spectrum

▶ Chasing the rainbow

▶ Wearing the halo

▶ Coloring the clouds

● ●

Have you ever noticed the side of the sky opposite the sunset? Next time everybody is oooing and ahhing over the blazing oranges and yellows and reds of the setting Sun, make a couple of sharp turns and check out the night side of the sky. Sometimes there is a blush of pink rose on its face. At first you might think it is reflected from the sunset, but there is no orange or yellow or red in this color. The color is violet, and behind it is a deep, dusky, purply blue. The sky on such an evening is like a great tub of colors that have become separated from one another through the day. All the yellows and reds are at one side, and all the blues and violets are at the other.

One of the sad things about pollution of the air over cities — aside from the fact that the people who live in them have to breathe the stuff — is how much fun it takes out of the sky. Take off from a busy airport and in a few minutes it hits you what you've been missing: Oh, yeah, *this* is a clear blue sky.

Up above the layer of stuff that so often hangs over a city, there's quite a show going on. This chapter is about all the things going on up there that you might be missing. Not the rain or the wind in your face, but the sky's beauties. It's about haloes and sun dogs and sun pillars and coronas and auroras and rainbows and glories. It's about all the colors of clouds and sky.

The colors are the work of Earth's atmosphere playing with light from the Sun. You may have noticed at night, without the special effects of atmosphere on sunlight, that it's a pretty black-and-white universe out there.

From dawn to dark, the sky is a light show brought to you by the atmosphere. Here's one for you: What if the air molecules didn't scatter the blue wavelengths? What color would the sky be if there was no scattering going on? The sky would be pitch black, and stars would be visible all day long.

Seeing the Light

The light of day is a mixture of different waves of energy radiating from the Sun. Each wave has a different length, as the next section describes, and each length is seen as a different color. When all the different waves strike about equally on the rods and cones in the back of your eye, you see white light.

The colors you see are the handiwork of gas molecules and other stuff in the air like water droplets and ice crystals in clouds, raindrops, and tiny bits of dust and material like ash and droplets of acid and other chemical compounds. Especially if you live in a city, you know this already: There's a lot of stuff up there.

The different stuff does different things to light. More than one of these effects may be at work at the same time. Here are the four main things going on:

- The sky is always *scattering* light waves. The tiny gas molecules are scattering short waves, making the sky blue. The clouds are scattering all wavelengths, which is what makes them white. Sunsets are red, orange, and yellow because these colors are scattered near the ground as the sunlight passes through the atmosphere at a sharp angle.

- A lot of bending, called *refraction,* is going on. Refraction is what makes a big part of rainbows. In fact, unless they are coming at you directly overhead, all light waves are bent by the atmosphere.

- Light waves are bounced back, or *reflected,* by ice crystals and raindrops. Special effects like haloes and sun dogs, among others, are the result of both refraction and reflection.

- Another kind of bending of light waves occurs when they pass around the edge of cloud droplets and raindrops. This *diffraction* causes the waves to bump into one another and, along with reflection, is what makes bright coronas around the Sun and moon.

Many special effects in the sky involve more than one of these processes. A rainbow, for example, is the result of light being bent one way by passing through raindrops, bent another way by passing around them, and also reflected off the backsides of raindrops. The colors of sunsets and sunrises also result from both the scattering and the bending of light waves.

It's a busy sky. All kinds of colorful things are going on, especially when clouds of different types are coming and going. Find yourself a good spot with a big view and check it out!

In Living Color

If somebody asked you what color the sunlight is, probably you would say it is white. I know I would. But the color white that you think of as sunlight really is an interesting and entertaining combination of all the colors that the sky has to offer. These are the colors of the spectrum, or arrangement, that you recognize most often as the colors of the rainbow. These colors are red, orange, yellow, green, blue, indigo, and violet.

While the combination of them appears as white coming from the Sun, each of these colors travels as a light wave of a different length. Red is the longest visible wavelength, followed in order of length by orange, yellow, green, blue, indigo, and violet. That is the *spectrum* — the order that you see them in when a rainbow brightens the sky. (There are *infrared* wavelengths that are too long to see, and *ultraviolet* wavelengths that are too short.)

The scattering and the bending of the sunlight and the other processes that separate the wavelengths into the various colors are happening all the time. The bright white light of the Sun over your head at high noon on a summer day is at that very moment somebody else's red and orange sunset. It's a matter of angle or slant. It's also a matter of the stuff in the sky between the sun and the color-seeing rods and cones in the backs of your eyeballs. But always it's the same order: red, orange, yellow, green, blue, indigo, and violet.

Why the Sky Is Blue

Be ready for this one. *Why is the sky blue?* It's on the tip of every little tongue that ever wagged, just waiting to make you feel like a nincompoop. You think she doesn't expect you to know the answer to this question? If you're wrong, you think she won't remember?

 The sky is blue because when light from the Sun comes down to Earth, it bounces off the tiniest parts of the air, and when this happens, the tiny parts of the light that make the color blue scatter out everywhere.

Because they are so small, the molecules of the gases oxygen and nitrogen, which make up most of the air, selectively scatter only the shortest wavelength rays of sunlight and allow the longer wavelength rays to pass on through. When you look up on a clear day, you are seeing those short wavelength blues that the air molecules are scattering in all directions.

Even on a day when the sky does not appear to be especially clear or blue — when it is clouded, for example, or not particularly clean — there's still a whole lot of scattering going on.

Reflecting on Clouds

You don't have to be a weather expert to pick up a thing or two about clouds. Without even thinking about it, everybody pretty much knows from experience when a cloud is just a big, billowy beauty to behold, when it begins to look like rain, and when it scares the daylights out of you. The shapes are important clues to what clouds are up to, of course, but what everybody really notices are the changing colors. In some areas where severe thunderstorms are part of everyday life, you can watch a cloud change in a matter of minutes from white to gray to black.

The tops of clouds are white because of their billions and billions of water droplets. Although they are tiny in comparison to raindrops, the droplets are large enough to scatter all wavelengths of sunlight. None of the color waves are being separated from the others. There is a whole lot of scattering going on. A large cloud will have a variety of shades of white and gray. It may have a bright white, cauliflower top, silvery gray sides, and a dark gray underbelly. The chances are small that a ray of sunlight entering a cloud will pass all of the way through it without bouncing off the surface of a water droplet.

As a cloud grows deeper, its bottom turns darker, and before long it begins to get everybody's attention. Two things can be happening. It has become so tall, there is so much scattering, that little sunlight is reaching the base of the cloud. At the same time, the water droplets inside the cloud can be growing larger near the bottom, and instead of reflecting sunlight off their surfaces, they are absorbing the light. The deeper the cloud, and the bigger its droplets, the darker its base. Less and less of the Sun's light is making it through.

Silver Linings

Sometimes light rays are bent when they travel around the edge of an object such as a water droplet while passing through a cloud. Depending on the size of the droplets and the gaps between them, this bending can cause the waves to interfere with one another, a process called *diffraction*.

When the Sun is shining from behind a dark cloud, this kind of light wave interference sometimes causes silver lining around clouds — the especially bright outlines of light you see around the edges of growing cumulus, for example. Silver linings are most often seen around clouds that contain especially large water droplets.

On rare occasions, this same slight bending of sunlight also can cause the undersides of middle-level *altostratus* or *altocumulus* clouds, which Chapter 5 describes, to shine with a splash of colors, an effect known as *iridescence*. This effect may be hard to spot because it usually occurs within 20 degrees of the Sun and is best observed with sunglasses.

Look for silver linings when clouds are growing in the east in the morning or the west in the afternoon — when the Sun is shining behind them. (See Figure 13-1.)

Figure 13-1: Silver lining around a growing cumulus cloud.

National Center for Atmospheric Research/University Corporation for Atmospheric Research/National Science Foundation.

Blue Haze

Look for blue haze when you have a long-distance view over heavily forested mountain valleys and canyons.

There's another blue in the air that has a different look than the clear blue sky. If you didn't know better, you might blame air pollution for what is called *blue haze* in many parts of the country. It is not the work of the same oxygen and nitrogen molecules whose scattering effects make the sky blue, but it's not pollution either.

The haze is caused by the buildup of natural organic particles in the air, and is most dense over heavily forested regions. The thinking is that hydrogen and carbon is boiled off in vapors from the natural processes of trees and other plants. In hot sunlight, they combine to form hydrocarbon compounds whose molecules scatter light. The haze appears blue because the hydrocarbon molecules are small enough to scatter the short wavelengths of blue.

Blue haze has been around a long time. It has given names to several regions of the United States. These include the Blue Mountains of Washington and Oregon, the Blue Ridge Mountains of Virginia, and the Great Smoky Mountains of Tennessee and North Carolina.

Sunbeams

Some cloudy afternoons when the air is full of haze or dust or tiny water droplets, the Sun will shine through a break in the clouds and put on a striking display. Rays of sunlight shine down from the sky in great spreading beams across the dark shadow of cloud. Sometimes it looks like a scene from a movie about the Bible. You expect lightning to flash and Moses to come down and part the Red Sea. The effect is especially striking when the rays shine through layers of haze.

In England, this event is known as *Jacob's ladder.* It also goes by the name *Buddha's fingers* or *Ropes of Maui* and *Sun drawing water* because sometimes it looks like the Sun is reaching down and drawing up water.

The same visual effect of rays radiating outward and upward from the Sun occasionally is seen when the Sun sets below the horizon through a hazy or dusty sky. Weather scientists have a particularly unfortunate name for this dramatic sight that makes it sound like some kind of distasteful ailment. They call it *crepuscular rays* — *crepuscular* having to do with twilight. But maybe you already knew that.

Look for sunbeams when the Sun is shining from behind a dark cloud, especially early or late on a hazy day.

Around the bend

Here's a mind-bending thought: The atmosphere of Earth allows you to see around the horizon.

Because the atmosphere bends the light of the Sun, especially at the horizon, you see its image in the sky something like four minutes longer every day than you would if there was no air.

The atmosphere bends the path of light around the horizon from the rising or setting moon or sun to your eyes. At the beginning of daylight and at the end, these images appear higher than they really are in the sky. So it appears to rise about two minutes earlier and sets about two minutes later.

Sunrise, Sunset

On a clear summer day, when the Sun is high in the sky, all wavelengths of its light are arriving at your eyes with about the same intensity. The sunshine is brightest and whitest about noon. The Sun had risen in the east in a red and golden glow, and all morning long as it climbed the sky, it faded from orange to yellow to the bright white of noon.

These changes in color are caused by the different amounts of atmosphere that the rays of sunlight must travel through at different times of day on their way to your eyes.

As its arc shifts lower in the sky through the afternoon, the rays of light strike through the atmosphere at a greater and greater angle. The process causes more and more of the light to both bend and scatter. The lower the Sun sinks toward the horizon, the more atmosphere its rays must travel through, and the more scattering of the short wavelengths of light by the air molecules. Finally, as the Sun nears the horizon, its rays travel through so much atmosphere on their way to your eye that all that is left is the long wavelengths — the yellows and oranges and reds of sunset.

Of course, one of the most striking sights of sunrise or sunset is the blazing colors it paints on the undersides of high clouds just before or just after the Sun rises above or sets below the horizon. People who make it a habit of watching sunsets look not so much for clear skies as they do for evenings when high clouds are likely to put on an especially colorful show.

Smoke from wildfires or clouds of human-caused pollution can fill the atmosphere with particles that scatter light and makes colorful sunsets and sunrises. Once in a while, the smoke or pollution becomes so thick in the afternoon that the red Sun fades from the sky even before it reaches the horizon.

Flattening the Sun

The bending of light by the atmosphere has some interesting effects you can check out anytime you get a good look at the horizon when the Sun is rising or setting. Notice that the Sun has a squished look to it just as it peeks up at sunrise or just before it falls from view at sunset. What is perfectly round most of the day at that moment looks like a bright red or orange raggedy edged biscuit.

The key to this odd shape is the fact that light waves are bent more by the atmosphere the closer they are to the horizon. They strike the layers of air at a greater angle and they have more air to travel through on their way to your eyes. The Sun's image is flattened because the bottom of the Sun is just a tad closer to the horizon than the top of the Sun.

Large clouds of ash and other substances from the eruptions of volcanoes can spread sunlight-scattering material far and wide. Even after the ash settles out of the sky, volcanic sulfur compounds and other molecules of gases can linger in the atmosphere for a long time. For example, colorful sunsets and twilight skies were especially noticeable around the world for many months following the gigantic eruption of the Mount Pinatubo volcano in the Philippines in 1991.

The same scattering and bending effects that make a colorful sunset often linger in the sky for quite some time after the Sun has set in the evening or before it rises in the morning. This is the time of *twilight*. In the summer in the middle latitudes where you live, twilight adds daylight to the morning before sunrise and to the evening after the Sun sets. In a world without atmosphere, there is no twilight. At sunset, the place becomes instantly dark.

The Green Flash

The dust and air molecules and other stuff floating around in the atmosphere cause light waves not only to scatter, but also to bend, a process known as *refraction*. The waves bend differently according to their lengths — the long reds the least, the blues and violets the most. As the Sun approaches the horizon, passing through the most amount of the Earth's atmosphere before disappearing, this bending has the effect of creating a vertical stack of wavelengths of the different colors.

Every once in awhile, especially as the Sun is setting over the ocean, conditions are just right to see a green flash of light on top of the big red Sun just as it sinks into the sea. The colors of the spectrum of sunlight's waves are disappearing one by one, although usually these features are lost in the scattering of the atmosphere. The short and easily scattered violets and blues are always lost to the eye, but that rare flash of green is a glimpse of color that only in the last second are not blocked out by the Sun's brightness.

Look for the rare green flash when you can see the sun set on the ocean.

Rainbows

Could there be a more wonderful sight in the sky than a rainbow? Well, I myself am not complaining, mind you, but I probably owe it to some angry people in muddy boots to make this one little observation: Rainbows would be just a tad more wonderful if at the end of these things there were in fact a pot of gold. (I had my people at the Go Figure Academy of Sciences check this out, by the way, so I can save you some time and a lot of trouble on that.)

When you are seeing a rainbow, you are looking at falling rain. (See Figure 13-2 and also the rainbow in the color section earmarked "Special Effects.") The Sun is at your back. Here's a little abracadabra for you:

Figure 13-2:
Rainbows
are simply
falling rain.

National Center for Atmospheric Research/University Corporation for Atmospheric Research/National Science Foundation.

- If you see a rainbow in the morning, you might begin looking for an umbrella.

- If you see a rainbow in the afternoon, you're in for a break of clear weather.

How do I know this? In the middle latitudes, where you live, weather travels from west to east. With the Sun at your back in the morning, you are facing the western sky, looking at weather that is heading your way. A rainbow in the afternoon means the western sky is clear.

Two interesting things happen to the sunlight that has passed over your head when it reaches the drops of falling rain. The sunlight is bent, or *refracted,* and also it is bounced back, or *reflected.* The white sunlight enters the rain-drop, reflects off the back side of the drop of water as if it were a mirror, and then goes back out the same side it entered.

But the different wavelengths in the light are bent when they enter the rain-drop because the water they are entering is more dense than the air they are leaving. The shorter wavelengths of light that are the colors violet and blue and green are bent more than the longer wavelengths of light that are the colors red and orange and yellow. The same thing happens when the light waves head back out of the raindrop after reflecting off of its curved back-side. The wavelengths bend again because the air they are entering is less dense than the water they are leaving. All of this bending causes the wave-lengths of visible sunlight — the different colors — to very clearly separate from one another.

What you are seeing in the rainbow is the different colors of sunlight spread out across the sky according to the lengths of their waves.

You are seeing only a single color from each drop of rain, because the reflected light waves enter your eyes from a different angle than the next. So it takes a bundle of raindrops falling to make a rainbow. And it is your very own personal rainbow, because even a person standing next to you is seeing the reflected light waves from different raindrops, at a slightly different angle.

The same bending and reflections of sunlight waves through falling water drops causes the colors of the rainbow you see occasionally in the spray of lawn sprinklers, fountains, and waterfalls.

Rainbows occur only when it is raining in one part of the sky and the Sun is shining in another, although it is possible to see the colors of the rainbow in the sky when it is not raining. This happens once in a while when the Sun is low in the sky, and ice crystals are in the air high overhead. The sunlight will reflect through the ice crystals and briefly you will see an arc of colors. Weather scientists give this effect an odd and extremely forgettable name. If you really want to know, they call it a *circumzenithal arc.*

Look for rainbows when rain is falling in one part of the sky and the Sun is shining at an angle at your back.

Once in a while, you will see a double rainbow in the sky. The first is bright and arced at the usual angle, and farther out a secondary rainbow will be visible. This rainbow is the result of sunlight reflecting twice off the backsides of raindrops and is definitely less bright than the primary rainbow. Notice something really interesting: Because it has reflected twice off the back of the raindrop, light that makes the secondary rainbow has its order of colors opposite a regular rainbow.

Haloes

A ring, or a *halo,* around the Sun or the moon is the most common of all special effects in the sky. Haloes are one of several effects caused by thin clouds of six-sided pencil-shaped ice crystals that reflect or bend the sunlight or moonlight that is shining through them.

The most common halo is a circle that is always the same size, covering 22 degrees of sky from the center of the moon or Sun to the rim of the ring. (Extend your arm and spread out your hand, and the distance from the tip of your thumb to your pinky covers roughly 22 degrees of sky.) The light strikes the crystals at a particular angle, some of it passing through the crystal and bending. You might notice a reddish glow on the inside and a bluish tint on the outside, although most haloes appear as mostly white light.

Once in a while a bigger halo, 46 degrees from center to edge, outlines the Sun or moon. This takes shape when the light shines through the end of the ice crystal and enters or leaves through one of the six sides.

Haloes around the Sun or moon were popular "weather signs" in ancient times. People thought a storm was coming when they saw a halo. Sometimes it's true that an advancing warm front will first appear as thin, high cirrus clouds of ice crystals that form haloes. Like all weather signs, however, haloes are not very reliable forecasts.

Look for a halo when high, cirrostratus clouds, which Chapter 5 describes, are in the sky.

Sun Dogs

A *sun dog* or mock sun is a odd sky critter that sometimes appears along the sides of a halo and sometimes shows up by itself. The sky can suddenly look like it contains two or three suns. Taking a closer look, however, usually you will notice blurred coloring effects on the smaller outer mock suns.

Sun dogs are the result of sunlight passing through flat platelike ice crystals that are falling through the thin cirrus cloud. They are falling like tiny flying saucers, so that the light strikes the flat top of the plate and gets bent as it comes out through one of the six sides of the crystal.

The color page earmarked "Special Effects" includes a photograph that shows a sun dog along the sides of a halo.

A similar effect to a sun dog appears at the top or the bottom of a halo once in a while. This spot of brightness, known as a *tangent arc,* is caused by the pencil-shaped ice crystals lining up and falling flat through the sky as if they were resting on the top of a desk. Haloes and sun dogs and similar special effects are most common in cold polar regions where the sky most often contains ice crystals.

Look for the rare sun dogs when the Sun is low in a sky that contains ice-crystal cirrus clouds.

Sun Pillars

When the Sun is low on the horizon, soon after rising or shortly before sunset, ice crystals in thin cirrus clouds can play another special trick. When they line up with their flat surfaces all facing upward as they fall, sunlight will bounce off the flat surfaces in reflection. This effect forms a bright pillar of light below or above the sun.

It's not just their shape that makes *sun pillars* different from similar effects. Pillars are different from sun dogs and coronas in their color — or lack of color. Sun dogs and coronas and even haloes will show the colors of a rainbow because their light is being bent as it passes through the ice crystals. Sun pillars are the same color as the Sun, because the light is being reflected off the surface of the crystal, as if it were bouncing off a mirror. The color page earmarked "Special Effects" includes a photograph that is a good example of a sun pillar.

Look for sun pillars at sunrise or sunset.

Coronas

A *corona* or crown can take shape around the Sun or the moon when its light passes through a cloud of tiny water droplets that are all about the same size. The light is bent slightly as it passes around the droplets and through the gaps between the droplets.

This is the kind of bending that weather scientists call *diffraction*. It causes the light waves to bump into one another. When this happens, sometimes the waves cancel each other out, making darkness, and sometimes they build on one another, making special brightness.

A corona can be bright white, or it can give the Sun or moon a crown of colors. A corona of colors around the Sun is sometimes called *Bishop's ring*. Coronas around the Sun are often more dramatic and colorful than those around the moon, but the brightness of the sunlight makes them harder to see.

This effect tells you something about the weather. When the bright disk of a corona appears, you know you are looking through water droplets rather than ice crystals. This means you are most likely looking at the Sun or moon through middle level altostratus or altocumulus clouds, which Chapter 5 describes, rather than the high cirrus clouds of ice crystals that make haloes and sun dogs.

When the cloud cover is broken, a misshapen but colorful corona effect can form. This is known as cloud *iridescence,* and the patches of color can be striking shades of the spectrum.

Look for coronas most commonly in the winter over mountains.

Glories

A *glory* is sort of a backward corona (see preceding section). Instead of seeing the light shining through a cloud of water droplets, the Sun is behind you, and you are seeing rings of color spread out from your shadow on the cloud.

A glory or brockton bow results when sunlight is bent as it enters and exits the water droplets and reflects back toward your eyes off the back side of the droplet.

The recipe for a glory sounds a lot like making a rainbow. The Sun is at your back, and its light is bending and reflecting as it passes through water. The big difference is that a glory is formed by the effects of light on tiny water droplets in a cloud. A rainbow is formed by the effects of the light on falling raindrops, which are huge in comparison to cloud droplets.

A glory is a common sight to airplane travelers. The shadow cast by the airplane on clouds below often is inside a circle of brightness and a ring of colors. Before airplanes, the effect was less commonly seen by people on mountains as their shadows were cast against nearby clouds. This effect was common in the Brockton Mountains of Germany, and it became known as a *brockton bow.*

Look for glories out the window of an airplane, where the plane's shadow is reflected on cloud tops.

Mirages

Seeing is not always believing. When air layers of dramatically different temperatures come together, light waves can wander all over the place. These temperature differences make for sharply different densities, and light moving in and out of these different densities changes the directions of lines of sight that your brain expects to be straight.

Things can appear higher or lower than they really are. They can appear upside down. Lakes and mountains can appear out of nowhere. It is not your imagination. It's not your mind bending, it's your light bending. It's the air playing tricks — mirages. Because they depend on extreme temperatures, mirages are most common in deserts and polar regions. Two kinds of mirages are common.

Look for mirages where you have long, distant views through especially hot or especially cold air.

Inferior mirage

When temperatures are especially hot near the ground, air is rising so quickly that it turns normal air densities upside down. Instead of being most dense near the ground, the air is "thicker" a few feet off a hot road or stretch of desert sand. This causes normally straight rays of light to bend upward, entering the eye from below. This makes things appear lower than they really are, and in some cases, even upside down.

This trick of the air is known as an *inferior mirage,* because things look lower. Light from the blue sky is bent in a way that makes it look like a puddle in the distance on a hot road, or a lake on the surface of the hot desert sand.

Superior mirage

When air near the ground is very much colder than air just above it, light rays bend in the other direction and things can appear taller than they really are. Light is bending down, entering the eye from a higher angle. Your eyes and your brain have no way of knowing that the light is not coming perfectly straight, so in a *superior mirage* objects look higher than they are. Across the snow fields of the Great Plains, for example, these *superior mirage* conditions can make the Rocky Mountains appear taller than they are.

Twinkle, Twinkle Little Air

While most special visual effects are caused by bending and reflecting sunlight and moonlight, Earth's atmosphere has other tricks up its sleeve.

One is the twinkling of a star. It's not really the star that is causing the twinkling. It's the air that is playing with that tiny beam of light from a nuclear fire millions and millions of miles away. It travels all that way in a fairly straight and true line until it hits Earth's atmosphere, and then it gets bounced around as it passes through layers of air of different densities. So entering your eye it flickers or twinkles, an effect that scientists call *scintillation.*

Planets usually don't seem to twinkle. This is because they are bigger objects, and their images are larger than the angles of flickering caused by the different layers of the air.

Another night sky trick affects both planets and stars and can be thought of as a superior mirage (see preceding section). Things appear higher in the sky than they really are. Their light is bent downward toward you as it enters the atmosphere, so unless they are directly overhead, they are not really as high in the sky as they appear. The farther toward the horizon, the greater the angle, the bigger the bend. More than this, I can't tell you. That is *Astronomy For Dummies* (IDG Books Worldwide, Inc.) by Stephen P. Maran, Ph.D.

Auroras

A curious and colorful glow of light high in the night sky is regularly visible only to people living in regions of the higher latitudes toward the poles. These lights are known as *auroras,* and their colors look more like something in a neon sign than a rainbow. Near the South Pole they are the southern lights, or the *aurora australis,* and near the North Pole they are the northern lights, the *aurora borealis.*

The Sun causes auroras, but their dancing light is not sunlight. It is the effect of an energy flow from the Sun known as the solar wind that meets the invisible lines of the magnetic field surrounding Earth's atmosphere. The North and South poles are like the ends of a big bar magnet, which makes compasses point the way they do.

Streaming from our favorite star is a flow of atomic particles known as electrons and ions that have been stripped of atoms of gases in the nuclear fire of the Sun. As these particles approach Earth they follow the magnetic field lines toward the poles and collide with atoms of oxygen and nitrogen gas high in the atmosphere, anywhere from 50 to roughly 500 miles up.

That is the light of the aurora — the effect of the solar wind hitting the upper atmosphere. Collisions with oxygen produces yellows and greens and reds. Collisions with atoms of nitrogen give off a blue tinge.

Look for auroras in the northern parts of the country on especially clear, moonless nights when you are away from the lights of a city.

This infrared satellite image shows Hurricane Andrew crossing the Florida coast and making landfall August 24, 1992, in Dade County, Florida.

The winds of Hurricane Bonnie that plowed into North Carolina in 1998 are bending trees to their breaking point. ©Jim Reed.

Hurricane Mitch, a Category 5 monster that killed more than 11,000 people in Honduras and Nicaragua, churns over Central America October 26, 1998, with winds of 195 miles per hour.

(Before Hugo)

(After Hugo)

Homes at Folly Beach, South Carolina, before — and after — Hurricane Hugo wiped them off the beach in September 1989.

10 NOV 97

JPL

Courtesy of NASA/JPL/Caltech.

NASA
cnes

The following images from the Topex Poseidon satellite reveal the powerful El Niño of 1997–1998 and the big La Niña that followed. The extra warm ocean temperatures of El Niño, shown in white, reached their peak in November 1997.

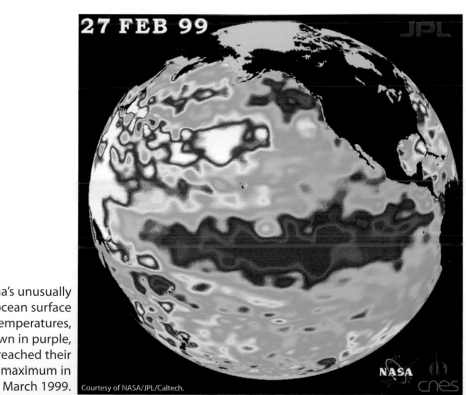

27 FEB 99

JPL

NASA
cnes

La Niña's unusually cool ocean surface temperatures, shown in purple, reached their maximum in March 1999.

Courtesy of NASA/JPL/Caltech.

Storm chasers document a severe thunderstorm in western Oklahoma.

©Jim Reed.

©Jim Reed.

Hailstones like these grapefruit-sized chunks in central Kansas damage airplanes and vehicles and have been known to kill livestock.

A memorial to one of the many lives lost in Moore, Oklahoma, in the historic outbreak of twisters May 3, 1999.

©Jim Reed.

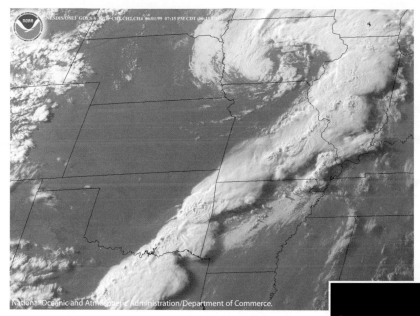

National Oceanic and Atmospheric Administration/Department of Commerce.

Squall-line thunder-storms ahead of a frontal system on June 1, 1999, stretched from Illinois southeastward over Missouri, northwest-ern Arkansas, south-eastern Oklahoma, and northeastern Texas. Such storms can produce large hail and tornadoes.

©Jim Reed.

National Center for Atmospheric Research/University Corporation for Atmospheric Research/National Science Foundation.

A tornado rakes across Texas on June 2, 1996.

A giant bolt of cloud-to-ground lightning fires up the twilight in south central Kansas.

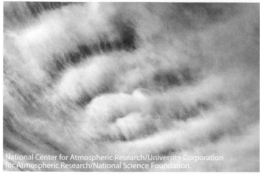

National Center for Atmospheric Research/University Corporation for Atmospheric Research/National Science Foundation.

Cirrostratus clouds.

Cirrocumulus clouds.

©Jim Reed.

National Center for Atmospheric Research/University Corporation for Atmospheric Research/National Science Foundation.

Cumulonimbus clouds.

Altostratus clouds.

National Center for Atmospheric Research/University Corporation for Atmospheric Research/National Science Foundation.

Nimbostratus clouds.

©Jim Reed.

Stratocumulus clouds.

Cirrus clouds.

Altocumulus clouds.

Stratus clouds.

Cumulus clouds.

Ice crystals in the sky form two special effects, a partial halo with parhelia, or sun dogs, on both sides of halo.

Photographer: Grant W. Goodge

Raindrops bend the rays of sunlight, separating its different wavelengths, or colors, forming a rainbow.

National Center for Atmospheric Research/ University Corporation for Atmospheric Research/National Science Foundation.

Photographer: Grant W. Goodge

Photographer: Doctor Yohsuke Kamide, Nagoya University. Collection of Dr. Herbert Kroehl, NGDC.

Most sun pillars are the effect of sunlight passing through ice crystals that are falling from a cloud.

High above the layer of sky where weather takes shape, Earth's magnetic field lights up the night sky near the poles with gusts of charged atomic particles flowing from the sun. Here, the Aurora borealis, or northern lights, shine over Anchorage, Alaska in 1977.

Chapter 14

Try This at Home

- -

In This Chapter

▶ Becoming weatherwise

▶ Following George, Tom, and Ben

▶ Making rainbows and clouds

▶ Bending the light

- -

*I*t's *your* weather, you know. Just because you can't control it doesn't mean it's not yours. (Hey, nobody else can control *theirs* either!) The rainbow, the cloud, the sunset — nobody else sees It from exactly the same angle you do. The snow and the rain that falls on your face and even the cool summer breeze that you feel on your arm — that's all yours.

You can take a personal interest in your weather, if you want to. A lot of famous and darn smart people have done just that over the years. In this chapter, I show you how to get started keeping track of your weather. I tell you about some of the great "amateur" weather-watchers in history and some of the great things they did. Also in this chapter are some cool experiments you can do to test some of the basic ideas about how weather works.

You're Not Just an Amateur

Amateur is a word that has fallen on hard times, especially in the sciences. Somebody pays a *professional,* while an amateur is, well, *just an amateur.* There's a good reason for this. After all, when I listen to a weather forecast, I want to know that this opinion about what it's going to be like tomorrow is coming from a pro. Still, *amateur* deserves more credit than it gets.

Weather science would not have gotten to first base without amateurs. The truth is, all of those professionals have built their weather science on instruments that were invented by amateurs, by Galileo and the boys 400 years ago. There was not a meteorologist among them. The science didn't exist. Mostly, they were just very smart people doing cool stuff at home.

Long before there were national weather services, keeping an eye on the weather was just something smart people did. Weather watching was an important part of exploration. When they arrived in the "new land" of North America, for example, early settlers from Europe realized very soon that just to survive the difficult winters they were going to have to keep their wits about them and their eyes on the sky.

Even now, while forecasters use very fancy supercomputers, still they rely on weatherwise amateurs. National weather services around the world use large networks of amateur weather observers. They get daily records of rainfall and temperatures from people in places that are hard to get to, and they depend on spotters for eyewitness accounts of tornadoes and other extreme weather events.

If you are keeping an eye on the sky, you are part of a tradition that is older than the United States. George Washington kept a weather diary even before the Revolutionary War. And Thomas Jefferson took time out from writing the Declaration of Independence to go get his first thermometer. And Benjamin Franklin was one of the most famous weather amateurs of all. When you watch the weather, no matter how you do it, you are following in some big footsteps, and you are following some good advice.

"Some people are weatherwise," Ben Franklin said, "but most are otherwise."

Galileo and the Boys

A lot of people have expressed a lot of opinions about the weather for ages and ages. It seems fair enough. Since everybody has to put up with the weather, everybody can have their say. For the longest time, listening to stories and opinions was about all there was to studying weather. A lot of interesting "weather lore" and proverbs have been passed down, but they are not what science is all about.

In science, to be able to say something is a fact, you have to be able to prove it in an experiment. And other people have to be able to copy your experiment and come up with the same result. No experiment, no agreement, no fact. So everybody needs to be using the same kinds of tools and measuring things the same way. But until the 1600s, there were no tools or instruments to measure weather's basic qualities: the air's temperature, its pressure, and its humidity, or the amount of moisture in it.

Galileo's thermometer

Credit for inventing the thermometer goes to the great Italian mathematician and philosopher Galileo Galilei. (This is the fellow who got into a lot of famous trouble with the Catholic Church for insisting that the Sun does not revolve around the Earth — that it's the other way around.)

In 1593, Galileo kept in his library a glass bulb about the size of a chicken egg with a clear thin tube sticking out of it so that he and his friends could watch the liquid inside the tube rise and fall as the temperature changed. The volume of the liquid expands when it warms up, he figured, and shrinks back when it cools.

Galileo had a lot of things going on — inventing the telescope, checking out sunspots, his teaching jobs, and the trial by the Inquisition. Meanwhile, other scientists picked up and ran with the *thermometer* idea. It took them a number of years to perfect the instrument. (It took the Catholic Church a number of years to admit that it was wrong about Galileo. The official apology came from Pope John Paul II in 1992.)

Torricelli's barometer

One of Galileo's students, Evangelista Torricelli, invented the barometer. He took a tube about four feet long, filled it with mercury, plugged one end to make a vacuum, and sank the open end in a pool of the liquid metal. He noticed that the mercury ran down the tube, but only so far, and then it abruptly stopped, always at the same spot on the tube.

Torricelli noticed that the level would rise and fall a little over time. He figured that the mercury in the tube was being held there by the air pressure — the weight of the air pressing down on the pool of mercury at the bottom of the tube. The small changes in the height of the mercury meant that the air pressure was rising and falling.

A French fellow, Blaise Pascal, figured out that weather changes could be causing the mercury in Torricelli's *barometer* to go up and down.

Cardinal de Cusa's hygrometer

People have always known that the air sometimes feels especially wet and sometimes feels especially dry, even if they didn't always know what to make of it. The first person to devise a way to measure the moisture in the air was a German fellow clear back in the 1400s, Cardinal Nicholas de Cusa.

Hang a big bunch of wool from one end of a rod, he said, and balance it with rocks at the other. The wool will absorb moisture from the air, and that end of the rod will go up or down depending on how wet or dry it is. It took a long time and many people to make an instrument that would accurately measure the humidity in the air, but de Cusa had the idea.

The most accurate *hygrometers* today work on a similar principle, except that they use human hair instead of wool. A strand of hair stretches as it absorbs moisture, and its change in length is measured on a dial.

Early American Weathermen

Even before the nation was born, the colonists from Europe were getting up close and personal with the weather. The English especially were very surprised — and disappointed — to learn that the winters in the New World were much more severe than the same seasons in Europe. The thinking at the time was that regions at the same latitude, or distance from the Equator, all had the same climate.

They didn't expect the cold that killed all but 32 of the original 105 colonists at Jamestown, Virginia, that first winter in 1607, and all but 50 of the original 102 pilgrims at Plymouth, Massachusetts, in 1620. And the pilgrims hadn't counted on hurricanes wrecking their supply ships.

Christopher Columbus was lucky he did not encounter a hurricane on his first round-trip in 1492. After all, a journey reaching land on October 12 would have crossed the Atlantic during the heart of the hurricane season. On later voyages, he did encounter hurricanes and was lucky again to survive. Some others who followed him were not so lucky.

The first careful, routine weather observations in North America were made in 1644 by a Lutheran minister, the Reverend John Campanius Holm, near what is now Wilmington, Delaware. Many early weather observers were ministers, and the theme of many sermons was how the latest storm was an act of God.

Personal weather watching was worthwhile and interesting business in the old days, and it still is. Some of the people who were busy putting the United States together as an independent nation took the time to watch and study what was going on in the sky over their heads.

George did it

George Washington kept a daily weather diary until December 13, 1799. His notations about the snow and the wind and the cold apparently were the last things he wrote. In fact, weather may have contributed to his death. He fell ill with a throat infection suddenly after riding around his farms in the cold weather he wrote about and died the next day.

Washington learned the importance of weather early in his career. In 1753, during his first military mission at the age of 21, the winter weather was so bad that it almost brought an early end to his military service — and his life.

More than once, General Washington used his skills as a weather forecaster to great advantage in his battles against the British during the Revolutionary War. On January 2, 1777, Washington found his troops trapped. On one side was mud, on another an icy river, and on another, the British. But a northwest wind was blowing, and Washington knew the mud would freeze. At midnight, his troops were able to retreat across the mud to safety. Again in the winter of 1778, Washington led a successful surprise nighttime attack against General Cornwallis' troops over muddy ground that had frozen in the chill of a northwest wind.

Tom did it

Thomas Jefferson took a break from writing the Declaration of Independence to go buy his first thermometer, and soon afterward he got himself a barometer. Jefferson collected weather information everywhere in the country he could find it. He exchanged letters with people far and wide about the weather in their regions. He kept a record of the weather around him every day for 50 years.

It was Jefferson who first realized the value of taking weather observations at different locations at the same time everyday. He exchanged letters with a fellow in Williamsburg, Virginia, and compared their 4 p.m. temperature, air pressure, and wind measurements over six weeks in 1778. What he discovered surprised him: The climate was cooler in his mountain home in Monticello than the Virginia lowlands of Williamsburg.

Jefferson was considered an expert on American weather and climate. He always argued that what the country needed was a network of weather observers. When he was president, the Lewis and Clark Expedition headed out on its long exploration of the Pacific Northwest, and Jefferson made sure that they took along accurate weather instruments.

Ben did it all

Benjamin Franklin, among other things, was a pioneer in important weather science. In his library, he had five thermometers and a barometer.

His most famous weather experiment — flying a kite in a thunderstorm to confirm that lightning was an electrical charge — was valuable but extremely risky business.

Don't try that at home. Franklin was lucky he had an insulator between himself and the kite string, but still it could have been fatal. Several people who copied Franklin's kite experiment were killed by lightning. Franklin's weather research led to his invention of the lightning rod, which Chapter 9 describes.

Franklin also was the first person to realize that the winds that make storms travel from one place to another are not the same winds that the storm blows in your face. Chapter 8 details his thinking about the storm that clouded over his plans to view an eclipse of the moon. Exchanging letters with his brother in Boston, Franklin figured out that although the storm was blowing winds *from* the northeast, other winds were carrying the storm itself *to* the northeast.

This idea takes a little getting used to. Feeling the northeast wind of a powerful winter nor'easter, for example, you might feel sure that the storm is coming at you from out of the northeast. But the fact is, as Franklin observed, nor'easters come up the East Coast from the south.

He was first to suggest that the eruptions of volcanoes affected the weather by cooling the atmosphere. He told scientists in England that he thought the unusually cool summer in Paris in 1783 was the result of an erupting volcano.

And Franklin was first to study the Gulf Stream, the warm ocean current in the North Atlantic that Chapter 4 describes, which is especially important to weather in the eastern United States and in Europe. Modern weather scientists are still studying the Gulf Stream.

Watching Your Weather

Watching weather is an even older tradition than talking about it, if you can imagine that. And it's just as much fun as it ever was. You can watch weather in a lot more ways now — more fancy instruments to read, more things to buy, and a load of stuff to check out on the Internet about the weather. But the fancy way is not always the best way when it comes to weather (and a lot of other stuff), and in fact, the most satisfying may be the least fancy approach.

Consider beginning with a weather diary, like the great founding fathers of the United States. Get in the habit of making a couple of readings on a shaded thermometer once in the morning and again at a regular time later in the day. Get a sense of the wind, its direction, and speed, and make a note of whether it is cloudy or clear. This time is also good to practice identifying cloud types. Diaries are cool anyway, and you're always free to add comments in each entry about weather conditions or any other darn thing that comes to mind.

Do you live in an area where volunteer storm spotters are used to help keep track of such things as tornadoes and other weather events that are hazardous to public safety? Check it out with your nearest National Weather Service office or your local emergency management agency.

Taking photographs is another interesting way to watch weather. There are two approaches to taking weather pictures. Some of your choice may depend on your camera equipment. One approach is to take pictures of big events like storms and especially interesting cloud formations, or lightning strikes, and special effects like rainbows and haloes and striking sunsets. Another approach is to record the small things about weather and the changes it takes during the seasons. The brilliant fall colors of leaves are old favorites in this category, but also small things like ice crystals on a window pane can make great photographs.

Getting Fancy

If you want to get fancy observing weather, you can get as fancy as your ambition will reach and your pocketbook will allow. You can observe the weather with the instruments and the techniques that experienced weather observers have always used, or you can take advantage of the electronic age and rig any number of gadgets to a home weather station. Fancy or not, no weatherwise home station is complete without an NOAA Weather Radio, which Chapter 1 describes, to provide vital information during weather emergencies.

Going instrumental

If you have the room in your yard, you can rig up an instrument shelter and keep more elaborate records. You can keep records for a period of time and put them in a spreadsheet or database program on your computer. And then you can compare your results with the records kept by nearby National Weather Service observing stations.

An instrument shelter (see Figure 14-1) is a big white box about four feet off the ground that has slatted sides to protect the instruments inside from direct sunlight but to allow air to move freely around the instruments. It has a double roof to prevent the Sun's radiation from heating up the box.

Figure 14-1:
An instrument shelter at a cooperative weather station in Granger, Utah, about 1930.

National Oceanic and Atmospheric Administration/
Department of Commerce.

The shelter contains these instruments:

- ✔ **A wet-bulb thermometer:** A wet-bulb thermometer is a well-ventilated thermometer surrounded in wet muslin that gives you the temperature of air that is saturated in water vapor.

- ✔ **A dry-bulb thermometer:** A dry bulb thermometer measures the actual air temperature.

- ✔ **A maximum temperature thermometer:** The maximum thermometer is designed so that it keeps track of the highest temperature reached that day.

- ✔ **A minimum temperature thermometer:** The minimum thermometer is rigged to mark how low the temperature fell.

- ✔ **An aneroid barometer:** The aneroid barometer is a more convenient mechanical version of a mercury barometer that measures air pressure on the face of a dial rather than heights of liquid in a tube.

The difference between the wet-bulb and dry-bulb thermometer readings allows you to figure out the relative humidity of the air.

Going digital

If you like electronic gadgetry, boy, did you come to the right hobby! All of the essential weather measuring instruments — as well as nonessential instruments — are available in digital gear. They range widely in precision and in price.

Instruments can be purchased as single-service components of a self-assembled home weather station. Others are designed and packaged as integrated home weather stations. Prices for electronic home weather stations range from under $250 to well over $2,000. Many come with consoles, which permit them to connect directly to a desktop computer, and have their own software for storing and graphically portraying information.

Other electronic devices and services allow you to keep up to the minute with changing weather conditions. You can be alerted by beeper or Internet-linked cellular telephone any time a storm alert is issued or a forecast is changed.

A wide range of computer software is available through software vendors and instrument retailers as well as archives over the Internet. The Appendix gives you a good idea of the huge number of weather-related resources available on the Internet. Computer users will find software available for any number of applications. They can fashion their own forecasts, for example, or use their systems to maintain constantly updated information about local weather conditions.

Cool Weather Experiments

A weather scientist by the name of Zbigniew Sorbjan at Marquette University in Milwaukee, Wisconsin, has written the book on the subject of cool weather experiments. The title of the book is *Hands-On Meteorology,* and it's published by the American Meteorological Society. Here are some well-known weather experiments you can try at home.

Making rainbows

Want to make a rainbow?

Technically speaking, you can't really make a rainbow unless you can make it rain. But I'm not going to get too technical about this. The colors of the rainbow are always hidden in the white sunlight unless something happens to make its different wavelengths separate from one another. As Chapter 13 describes, the different wavelengths appear to your eyes as different colors.

Go Figure Academy of Sciences bites the dust

I was planning on having my people at the Go Figure Academy of Sciences come up with some really cool experiments about the weather. It was going to be great. You know, lightning, thunder, the works. But that's not going to be possible now.

I kind of lost control of my people after I had them try to chase down the end of a really nifty-looking rainbow in New Jersey to check out the pot-of-gold idea. They all came back muddy and mad at me. You might say the climate of the place completely changed. Still, I figured I owed it to you, to make sure about the gold. And besides, what if there had been gold? You know, running an academy of sciences is expensive, even one I made up, and I could have used the dough.

Anyway, to make a long story short, the whole ungrateful bunch of them has gone back to the Massachusetts Institute of Technology. Go figure.

With a spray from a garden hose, you can come pretty darn close to copying what it takes to make one of the most beautiful effects in nature. You've got the basic ingredients for a rainbow — the Sun at your back, and the falling drops of water in front of you. Make sure that the spray is below your eye level. The colors will show up in a semicircle band in the stream of water. Now change the spray coming out of the nozzle so that it comes out in bigger and smaller droplets of water. The rainbow changes brightness and width just like it does as sunlight is bent by different sizes of falling raindrops.

Here's an easy one to try inside. Fill up a jar with water and place it on the window sill so that the Sun shines down through it. Then place a sheet of white paper on the floor so that the light passing through the jar shines on the paper. The water jar should bend the different lengths of light waves as they pass through the jar and spread them out in the colors of the rainbow on the sheet of paper.

Bending light by refraction

Want to bend some light?

It's easy to see light bend as it passes through materials of different density. Turn the lights out and shine a flashlight into a glass of water. If you point the beam straight down into the water, its path is straight. But shine it at an angle and watch the light rays bend downward in the glass. The light waves are changing speed as they go from the air into the thicker, denser liquid. And the speed change is what gives them the bends.

Here's another light-bending trick with a glass of water. Turn the lights on and stick a pencil in the water. Move the pencil closer to you and then farther away from you and watch the submerged image of the pencil become thicker and thinner. That's the curved shape of the glass changing the amount of water the image of the pencil is traveling through, causing the light waves to spread out one way and come together another.

Bending light by diffraction

Want to see what it looks like when light waves bump into each other?

Here's an optical effect at your fingertips. Well, more exactly, between your fingers. Look through the beefy parts of your first and second fingers, letting just enough light through to see, and you will notice dark lines running through the image you see. It looks as though you are seeing the world through a set of window blinds. This is the effect of light bending through the tiny space on the way to your eyes, just as it does when it passes through the gaps between water droplets to form coronas, or between raindrops as part of rainbows. The dark bands are where the wavelengths of light interfere with one another in a way that they cancel each other out.

Weighing in on air

Want to prove that air has weight?

You and I know that air has *volume* — takes up space — without even thinking about it. That's what balloons are about — air taking up space inside them. And cars move around on cushions of air that take up space inside rubber tires.

To show that air has weight, you need two balloons, a rod, and some string. Blow up each balloon to the same size so that each contains the same amount of compressed air. Then tie them to each end of the rod with the same length of string. Now attach a length of string to the middle of the rod and adjust it so that the rod is straight across, so that the balloons are evenly balanced. Now pop one of the balloons and watch what happens to the rod. It will droop down on the end that holds the inflated balloon.

This is similar to the experiment that Galileo Galilei did in 1638. He pumped a lot of air into a glass flask and weighed the flask very carefully before and afterward. When the compressed air was in it, the flask was heavier. But many people laughed at Galileo about this for years afterward. Any fool could see that air had nothing in it to weigh!

Of course, proving that air has weight is not the same thing as finding out the weight of air. That took some fancy figuring many years later with some sensitive instruments to get just right. So, how much does air weigh? At sea level, a cubic yard of air — about 202 gallons of the stuff — weighs about 2 pounds.

Testing the pressure

Here's a way to get the idea of how air pressure changes with height, although this experiment involves water.

Take a soda can or some other fairly good size container that you can punch holes into. With something like a nail, put one hole maybe a third of the way down from the rim of the can and another whole quite a ways closer to the bottom of the can. Now fill it with water and look at the different shapes of the spouts coming from the holes. The top spout droops down more quickly than the spout coming out of the bottom hole, because the bottom has all the pressure of the water in the can pushing out, while the top hole has only the pressure of the water above it.

Now take another can and punch a few holes all at the same level, like down near the bottom. Pour water into this can and watch the spouts. Because they all have the same amount of water above them, they have the same pressure. So the spouts shoot out at the same curves and the same distance.

The greenhouse effect

Want to see what the "greenhouse effect" is about?

You need two thermometers mounted on two pieces of cardboard in a way that allows them to stand upright. Stand them side by side in a quiet sunny spot with their cardboard backs to the direct sunlight. Now place a large glass jar over one of the thermometers. Before long, you will notice that the thermometer in the jar measures a warmer temperature. The jar is acting like a greenhouse, preventing the warm air from rising away as it does around the other thermometer.

This is exactly like a greenhouse that a nursery uses. But it is *not exactly* the greenhouse effect that weather scientists have in mind when they talk about the atmosphere. A nursery's greenhouse roof prevents air from circulating, for example, while under the atmosphere's greenhouse gases, the whole world of weather is going on.

What weather scientists mean by the greenhouse effect is this: Some of the warmth from the Sun that is radiating up from the Earth in the form of invisible infrared rays is absorbed by these greenhouse gases in the atmosphere. So instead of escaping out into space, some of rays warm the atmosphere and some get sent back toward Earth. Like the greenhouse, the effect is that Earth is much warmer because of these gases in its atmosphere than it would be without them. (And a good thing too, by the way. Without this natural greenhouse effect, you and I would freeze to death down here!)

Cloud in a can

How about making a cloud?

You'll need to fashion a device called a *cold chamber.* With a contraption not too much different from this, scientists first experimented with seeding clouds to produce rain.

You will need two cans, one very large (or a pail) and one smaller. Put some ice mixed with salt in the bottom of the larger can. Place the small can on the ice so that its top is level with the top of the larger can. Pack ice and salt in the space between the cans. The large can will grow very cold, so it's a good idea to wrap a towel around it to protect your hands. Soon the air in the small can will become very cold. When you breathe into the small can, the cold will condense the water vapor in your breath into a cloud.

A bottle of fog

How about making a bottle of fog?

Fill a big jar with hot water, and then pour all of it out except for about an inch of water in the bottom. Now put a strainer over the top of the jar and put a bunch of ice cubes in the strainer. In a few minutes, the cold air from the ice cubes will sink down, and the air rising from the warm water in bottom will cause condensation to take place, just like Chapter 5 describes. What you have is a bottle of cloud, or fog.

Part V
The Part of Tens

The 5th Wave
By Rich Tennant

"Take it easy everyone. Let's just hope the wind currents carry this thing out of here."

In this part . . .

Here is the Weather Hall of Fame, a collection of weather disasters of the century around the United States and around the world. Here are the worst floods, the biggest hurricanes, the most severe winter storms, and the driest droughts.

Also in this part is another slant on weather history. There was weather to worry about long before there were any good explanations for it. Some of the best of the really old explanations have been passed down through the generations as weather lore, a collection of sayings and proverbs and signs that seemed to help predict the weather. Check them out for the fun of it and take them with a grain of salt!

Chapter 15

Ten (or So) Biggest U.S. Weather Disasters of the 20th Century

In This Chapter

▶ Checking out the worst of bad U.S. weather

As the 20th century was coming to an end, leading weather scientists found themselves scratching their heads over an unusual problem. They were being asked to list the biggest weather and climate events of the past 100 years. At times like these, many people have the urge to look back at major historical events to get a sense of how things have been and where they seem to be going. Or maybe it's just a media thing. Anyway, people try to give rank or special order to the importance of events of all kinds when things like centuries come to a close, but often it doesn't work every well.

As you can see in this chapter, weather events are just too different from one another to be compared that way. How do you compare a flood with a drought? A tornado with a blizzard or a hurricane? Even comparing storms and other events of the same type doesn't work as well as you might think. There are statistics to rank them by, but they don't help too much. A Category 5 hurricane sounds more important than a Category 3 hurricane, for example, and certainly its winds are stronger. But the flooding rains of a 3 may result in more damage and deaths than the 5. Do you rank events according to the number of lives they claim? According to the property damage they cause? In the end, no single way of looking at the great weather disasters seems quite right.

Dozens of scientists at the National Oceanic and Atmospheric Administration looked back over the weather and climate events of the 20th century and came up with 15 that most everybody agreed were the tops for one reason or another. They stood out because they were especially unusual weather events or because they powerfully changed people's lives.

In no particular order, here are what many of the nation's top scientists agreed were the top weather and climate events of the 20th century.

The Galveston Hurricane

On September 8, 1900, one of the most powerful hurricanes of the century came as a surprise to the people of Galveston, Texas. Weather forecasters knew there was a hurricane out in the Gulf of Mexico, but they didn't know where it was. This may seem strange in an age of satellites and "hurricane hunter" aircraft. But in those days, tracking of a storm at sea was hit or miss.

A storm surge 20 feet high came up onto the sandy barrier island that the bustling Gulf Coast city is built on and completely swamped Galveston. All through the night, the hurricane's winds whipped around the huge rafts of floating wreckage as hundreds of people hung on for their lives. Galveston weather service officer Isaac Cline said that the next morning revealed "one of the most horrible sights that ever a civilized people looked upon."

More than 8,000 people were killed. The exact death toll will never be known, officials said, because whole families were wiped out. The Galveston Hurricane was the deadliest natural disaster in U.S. history. A recent popular book, *Isaac's Storm* by Erik Larson (Crown Publishers, 1999), is all about the Galveston Hurricane and forecaster Isaac Cline.

The Dust Bowl

The climate event known as The Dust Bowl lasted for ten years — the Great Depression decade of the 1930s — and set many heat records that still stand across many parts of the Plains. It turned 100 million acres into dust. Awful dust storms known as *black rollers* or *black blizzards* roared across the Plains. The sky would darken for days. Research by weather scientists shows that the drought came in four waves: in 1930–'31, 1934, 1936, and 1939–'40.

Coming as it did during the Great Depression, the long drought drove hundreds of thousands of families into a desperate migration from the Plains to southern California and other states. The great agricultural losses deepened the nation's economic crisis and worsened the misery. The Dust Bowl experience alerted the country to watch out for long-term climate shifts as well as short-term weather extremes.

Super Tornado Outbreak, 1974

On April 3, 1974, a strong storm system crossed through the nation's midsection with a set of conditions that was perfect for breeding tornadoes. Before it was over 17 hours later, on April 4, a total of 148 tornadoes had roared across 13 states. The death and destruction ranged from the Great Lakes southward through the Ohio and Tennessee valleys and into Mississippi and Alabama. Among the twisters, 48 of them were killers, claiming a total of 315 lives. Hardest hit was the little town of Xenia, Ohio, where in a matter of minutes an F-5 tornado killed 33 people, injured 1,600 others, and destroyed 1,300 buildings.

Six of the tornadoes of the super outbreak were in the highest F-5 category, and scores of others were F-4 tornadoes. For all the damage and destruction, tornado specialists credit this swarm with giving them some important advances in their science. This outbreak gave them their first close look at microbursts, for example, which proved to be so dangerous to airplanes. For more on microbursts, see Chapter 8.

Hurricane Camille

A small but incredibly powerful 1969 hurricane, Camille was a Category 5 with wind gusts estimated at nearly 200 miles per hour. Camille came roaring out of the Gulf of Mexico and into Pass Christian, Mississippi, on August 17. The storm surge was measured at 24.6 feet, so high that one survivor was washed clear over the town without hitting utility poles, trees, or buildings.

The storm moved inland and turned toward the northeast, plowing into Tennessee, Kentucky, and Virginia with torrential rains of as much as 31 inches in central Virginia. Camille killed 256 people and destroyed more than 5,000 homes. Among the dead were 113 people in Virginia, a coastal state that saw its worst natural disaster come not from a hurricane out of the Atlantic but from the inland west.

The Great Midwest Flood

Heavy rains in the Upper Midwest through the spring and summer of 1993 led to the costliest floods in U.S. history. They began in June and lasted into August. In fact, some areas along the Upper Mississippi River remained above flood stage for five months. More than 15 million acres across nine states were flooded. Whole towns were under water. More than 54,000 people were evacuated from their homes. The water damaged 22,000 houses, claimed 48 lives, and did more than $18 billion in damage.

While most previous flooding along the Mississippi and elsewhere prompted calls for more dams and levees, the floods of 1993 were so big that officials began rethinking that idea. Instead, several efforts were made to change land use practices and move people out of the big floodplain.

El Niño Episodes

The Pacific Ocean climate condition known as El Niño developed two especially powerful episodes toward the end of the 20th century, in the early 1980s and the late 1990s. The unusual warming across the tropical Pacific shifted weather patterns around the world. Severe droughts caused famines and fires in some parts of the world. Especially strong storms brought flooding that caused billions of dollars in damage in the U.S. and elsewhere.

The first big El Niño, in 1982–'83, caught weather scientists by surprise. The second, in 1997–'98, was detected months in advance. Because they saw these conditions taking shape in the Pacific, National Weather Service forecasters were able to accurately forecast six months in advance the kind of winter the U.S. could expect in 1997. Such long-range forecasts are not always so accurate.

Hurricane Andrew, 1992

The costliest hurricane in U.S. history struck southern Florida and Louisiana in late August 1992. Hurricane Andrew missed the Miami area by several miles, but leveled 138,000 houses and other structures. It left 250,000 people homeless and caused damage estimated at $25 billion. Andrew was blamed for 23 deaths, but widespread warnings undoubtedly saved many lives. The heavy damage alerted communities in the Southeast to the need to bolster their building codes.

Forecasters accurately predicted that the hurricane would come ashore south of Miami and warned three days in advance — inaccurately, as it happened — that its track would come dangerously close to New Orleans. The hurricane actually struck a sparsely populated section of the south-central Louisiana coast.

New England Hurricane, 1938

The worst weather disaster ever to strike New England roared ashore on September 21, 1938. It smashed into Long Island, New York, and Providence, Rhode Island, with a giant storm surge that arrived just at the time of high tide. The center of the storm passed west of Providence. This meant that the storm surge and the highest winds hit the city directly from the south. Downtown Providence was under 20 feet of water. Wind gusts hit as high as

186 miles per hour at the Blue Hill Observatory in Milton, Massachusetts. The death toll was nearly 600, and property damage was estimated at $400 million. The hurricane missed downtown New York City by only 55 miles. Because it passed to the east, its winds were from the north, a circumstance that helped reduce damage from the surge of sea water.

This first tropical storm to hit New England in modern times caught weather forecasters completely by surprise. They had predicted only showers and breezes for that day.

Superstorm, March 1993

One of the worst snowstorms to hit the eastern third of the U.S. swept from Georgia to New England when everyone was thinking of spring. The Superstorm of 1993 struck in the middle of March. More than 40 inches of snow fell in seven states. It closed every major airport from Baltimore to Boston and dumped record snowfalls in Georgia, Tennessee, and the Appalachian Mountains. There were snowdrifts more than ten feet deep in Tennessee and Pennsylvania. Tornadoes struck Florida along with hurricane-force winds and a deadly storm surge that swept 11 people to their deaths along Florida's Gulf Coast. In all, 270 people died, and damage totaled $3 billion.

The Superstorm was accurately predicted by commercial weather services and the National Weather Service well in advance. In fact, the computer model operated by the European Center for Medium-Range Weather Forecasts saw this storm coming a week ahead of time.

Tri-State Tornado, 1925

The most deadly tornado in U.S. history hit the Ohio Valley on March 18, 1925. It traveled 219 miles across southeastern Missouri, southern Illinois, and southwestern Indiana. This single tornado killed 695 people, including 234 in the little town of Murphysboro, Illinois. Moving more than 60 miles an hour at times, the giant twister stayed on the ground three and a half hours.

In 1925, the National Weather Service did not issue tornado warnings, and its scientists did not know very much about them. The Tri-State Tornado alerted researchers to the idea of a tornado season and paved the way for research that led to official tornado warnings, which finally began more than 20 years later.

Tornado Outbreak of May 1999

The most costly outbreak of tornadoes in U.S. history caused $1 billion in damage in less than 48 hours on May 3-4, 1999, in Oklahoma and Kansas. At least 74 twisters touched down, including an F-5 monster that raked the southern outskirts of Oklahoma City. Doppler radar at one point measured winds of 318 miles per hour, the strongest ever recorded. Forty-two people were killed, and 800 were injured.

Using new technology, the National Weather Service was able to issue warnings during this outbreak that were an average of 21 minutes in advance. (The average "lead time" is 11 minutes.) In some areas, residents were warned 30 minutes before a tornado struck.

The Great Okeechobee Flood and Hurricane of 1928

One of the most deadly weather disasters in U.S. history was a huge Category 4 hurricane that struck Florida on September 16, 1928, with winds of 150 miles per hour, torrential rains, and flooding.

The U.S. death toll was 1,836 in Florida, including many agricultural workers who lived near the southern shore of Lake Okeechobee, in the Everglades region of south-central Florida. The big storm whipped up 15-foot waves in the shallow water and blew water out of the lake, causing widespread flooding that washed away entire communities.

This hurricane killed another 1,575 people in the Caribbean, although some estimates put the total number of dead at 3,500. In Puerto Rico, where at least 300 people died, the storm left 200,000 people homeless.

Florida Keys Hurricane, 1935

The first Category 5 hurricane to strike the United States in modern times was the 1935 Labor Day Hurricane, which swept over the Florida Keys. (The second Category 5 was Camille in 1969.) With winds estimated at 200 miles per hour, this hurricane goes on record as the most powerful ever to strike the U.S. coastline. The hurricane developed in the Bahamas, due west of Florida, and just a day before it struck, forecasters and residents had no idea of the power of the storm that was on the way.

A train was sent from Miami to rescue several hundred World War I veterans working on a highway project on the Keys. As the desperate workers were scrambling aboard at the station on Islamorada, the hurricane sent a storm surge 17 feet high over the island, sweeping the train from the tracks and drowning the workers. The death toll was more than 400.

New England Blizzard, 1978

Southern New England, especially Rhode Island and Massachusetts, was brought to a standstill for a week by a storm in 1978 that brought hurricane-force winds, coastal flooding, and blizzard conditions. Beginning on February 6, the region began feeling the impact of the collision of three air masses, from western Pennsylvania, northern Georgia, and moist marine air from the Atlantic offshore of Cape Hatteras, North Carolina.

Before the storm was over on the evening of February 7, as much as 4 feet of snow was on the ground across Rhode Island and the Boston area measured up to 38 inches of snow. The death toll was put at 17, and thousands of people were stranded far from home for several days.

Storm of the Century, 1950

Across the Midwest and the East Coast, an unusual and violent combination of weather conditions took shape during the Thanksgiving Day holiday in 1950. Warm, moist air was flowing down from the north, and cold air was flowing from the south. Taking shape over North Carolina, the storm moved up into Pennsylvania, northwest over Lake Erie, west, and then south and then east over Ohio.

Dubbed the "Storm of the Century" at the time, it brought hurricane-force winds, heavy snow and torrential rains to 22 states, and widespread coastal flooding from New Jersey through New England. It killed 383 people and caused property damage of $70 million.

The storm became a test case for National Weather Service scientists who were just then developing the first computer models for forecasting weather. They used data from the storm to try out numerical experiments to predict the large-scale circulation associated with mid-latitude winter storms.

Chapter 16

Ten (or So) Worst World Weather Disasters of the 20th Century

In This Chapter

▶ Seeing how the rest of the world lived

The weather and climate of the 20th century was not very kind to many parts of the world. Except for the Dust Bowl years of the 1930s, the U.S. was focusing mainly on the hammering it took from fairly short-term, dramatic weather events — damage for this storm or that. In many parts of the rest of the world, some of the worst disasters of the 1900s were the grinding, long-term effects of shifting climate. A flood can come from a rainstorm, a single weather event. Drought, or lack of water, on the other hand, represents the failure of a season or more.

Drought and its consequences on humans — famine and disease — was far and away the biggest killer of the 20th century. The hard economic reality of the rich and poor made itself felt, especially in the early years of the century. When a wealthy nation suffers a harvest setback, it can buy its way out of the food shortage on the world market. In a poor nation, when the rains fail, life fails. In the poor and populous nations of Asia and Africa, the losses of life were staggering. While the U.S. has a well-deserved reputation for violent weather, the death toll from weather and climate events elsewhere in the world is measured on a whole different scale.

The second biggest killer was drought's opposite — major river flooding from torrential rains and its terrible impact on health and food, on disease and starvation. Especially in the Yangzte Valley of central China, millions of lives were lost to the flooding of the giant river. Many of the worst floods accompanied typhoons and cyclones — the words for hurricanes in Asia and the Southern Hemisphere.

As they did with U.S. weather events (see Chapter 15), the experts at the National Oceanic and Atmospheric Administration looked over the biggest floods, typhoons, hurricanes, droughts, heat waves, tornadoes, winter storms, and climate events around the world. As the 20th century came to a close, they came up with this list.

Droughts

While the death tolls and other effects from storms are fairly obvious, the impacts of the failure of the rainy seasons can drag out over several years and are not so easy to track. The numbers of deaths related to the effects of drought are all estimates. Still, the really big drought, famine, and heat wave disasters would be hard to miss.

India

The drought of 1900 brought starvation and disease that killed up to 3.25 million people. In 1965–'67, more than 1.5 million perished from drought.

China

More than 24 million people died from starvation in the Chinese Famine of 1907. In 1928–'30, more than 3 million died in northwestern China. In 1936, another famine claimed the lives of 5 million. And in a drought in 1941–'42, another 3 million starved.

Soviet Union

In 1921–'22, drought claimed as many as 5 million lives in the Ukraine and Volga regions of the former Soviet Union.

Africa

Several devastating droughts struck the region of Africa south of the Sahara Desert known as the Sahel. There were major droughts there from 1910 to 1914, from 1940 to 1944, and a long and brutal dry spell from 1970 to 1985. The Sahel drought killed more than 600,000 people in 1972–'75 and another 600,000 or more in 1984–'85.

Floods

Heavy rains in Asia caused the greatest flooding disasters of the 20th century, especially in the middle and lower reaches of the major rivers in China.

China

The Yangtze River Valley was the site of disastrous flooding in 1900, 1911, 1915, 1931, 1935, 1950, 1954, 1959, 1991, and 1998. The worst, the Yangtze River Flood in the summer of 1931, came after the drought of 1928–'30 had killed more than 3 million people. In July and August, floods from torrential rains killed 3.7 million people by drowning, disease, and starvation. The flood affected 51 million people — a quarter of China's population at the time.

Vietnam

In 1971, heavy rains in northern Vietnam caused severe flooding that killed an estimated 100,000 people.

Iran

In 1954, the Great Iran Flood, caused by a single storm, claimed more than 10,000 lives.

Typhoons, Cyclones, and Hurricanes

While improved storm forecasting, advance public warning, and timely evacuations have sharply reduced the death tolls of hurricanes in the U.S., the same storms continue to claim thousands of lives elsewhere in the world. (The giant tropical storms called hurricanes in the Western Hemisphere are called typhoons in the western Pacific and cyclones in the Indian Ocean and the Southern Hemisphere.) The region of the western Pacific Ocean known as Typhoon Alley includes the coasts of China, Korea, Japan, the Philippines, and Southeast Asia, which have been hit with devastating storms.

Bangladesh

In November 1970, the greatest tropical storm disaster of the 20th century struck the low-lying coastal region of the south Asian nation of Bangladesh when a cyclone swept up out of the Indian Ocean. Powerful winds and a giant storm surge swept between 300,000 and 500,000 people to their deaths.

In 1991, another cyclone struck the country with 150-mile-per-hour winds and a 20-foot storm surge. More than 138,000 people were killed.

China

The coast of eastern China was struck by several large typhoons in the first half of the 20th century, causing terrible hardship and deaths that numbered in the tens of thousands. Among the worst was a storm in 1912 that killed 50,000 people along China's Pacific coast and a typhoon in 1922 that claimed another 60,000 lives.

Honduras

Central America fell victim in November 1998 to one of the strongest late season hurricanes ever known to form in the western Caribbean. Hurricane Mitch moved slowly through the warm waters and stalled for days over the mountainous regions of Honduras and Nicaragua, where it dumped rainfall measured as high as 75 inches. The floods wiped out the bridges and roads and other public facilities in Honduras and tore away whole villages in Honduras and neighboring nations. The death toll was estimated at 11,000.

Japan

In September 1958, Typhoon Vera dealt Japan its worst storm disaster. Nearly 5,000 people were killed, and 1.5 million were made homeless. Landslides, floods, and winds of the storm wrecked the nation's economy, wiping out roads, bridges, and communications systems.

Philippines

Typhoon Thelma brought terrible landslides and flash flooding to the Philippines in October 1991. An estimated 6,000 people died, many on Leyte Island, where logging had stripped vegetation from the land. This led to terrible landslides.

Winter Storms

Terrible winter storms brought death and destruction to the world more than once in the 20th century.

Iran

In February 1972, four years of drought ended in Iran with a week-long blizzard that killed about 4,000 people.

Europe

Savage winter storms in January and February 1953 drove storm surges onto coastal regions of the Netherlands and the United Kingdom. Almost 2,000 people drowned in the flooding.

Pollution

Once in a while, weather conditions gang up with air pollution to produce truly terrible events. This is what happened in the little steel town of Donora, Pennsylvania, in 1948, and in London, England, more than once. For more detail on these events, see the sidebars in Chapter 11.

Donora

The steel mill, zinc smelter, and sulfuric acid plants were pumping their stuff into the air over the little Pennsylvania town of Donora in October 1948 just like they had for years. But the weather changed, and nobody understood what it meant until it was too late. A high pressure system settled over the region. Winds died and a dense fog combined with the pollution to brew a dark and deadly smog. It killed 22 people and left thousands sick.

London

Weather and industrial pollution combined in December 1952 to cause an event known as the Great Smog of London, by far the worse air pollution episode of the century. Stagnant weather conditions, including fog, caused the air in the city to reach poisonous levels, killing 4,000 persons and contributing to the deaths of another 4,000.

Chapter 17

Ten Crafty Critters

People have wanted to know what the weather was going to be like long before there was any really good way of finding out. Weather was something people have *always* wondered about, in fact, and before there was weather science, they looked for ways to figure it out. Just imagine, the poor dears. no clock radio waking them up with a forecast, no last minute Weather Channel summary, and nothing on the subject in the newspaper. What was a person to do?

There were lots of ways to look at the weather before there was science. For a long time, in many parts of the world, the weather was said to be the work of the gods. If weather was good, the gods were happy, and when it was bad, they were angry — usually because of something the likes of you and I had done wrong.

Farmers and sailors came up with their own ways of looking at the sky and predicting what was going to happen next. Some of the old proverbs that Chapter 18 describes come down from this tradition. People looked for signs of changing weather everywhere — in the stars, the moon and Sun, the clouds in the sky, of course, and in the behavior of animals around them.

This chapter takes a look at this very old tradition of hand-me-down weather signs from the critters that share the house, the barnyard, and the wilderness with the likes of you and me. The idea was that these "lower forms" of animal life were more sensitive to weather changes than brutes like you and me. There may be something to that idea, in a way, but to be honest, as accurate predictors of weather, these "signs" don't hold up too well under close scrutiny.

So don't start bothering your cat for a weather forecast. Read these for the fun of it!

Cats

Some of the weirdest old animal signs of weather change have to do with cats around the house. My guess is they were thought up and passed around by people who spent too much time indoors.

When cats sneeze, for example, it was considered a sign of rain. It was also a sign of rain when a cat scratches itself or scratches a log or a tree. If a cat lies on its head with its mouth turned up, that meant a storm was on the way. A cat washing her head behind the ear was considered a sign of rain, although in other places, cats washing themselves were considered signs of fair weather.

Go figure. I don't know about you, but if there is a bigger mystery in the world than tomorrow's weather, it is understanding the behavior of my cats!

Dogs

Dogs haven't done as well as cats in the weather prediction business. At least, that's my explanation for the fact that there aren't very many old weather signs around that give much forecasting credit to dogs. Like most animal weather signs, the behavior of dogs most often was described as signs of rain.

A dog digging a deep hole in the ground was a sign of rain, for example, and so was a dog eating grass in the morning. A dog howling when someone leaves the house was said to be a sign of rain, although, for the life of me, I don't know why it wasn't a sign that the dog wanted to go with them. A spaniel sleeping was considered a sign of rain. Imagine that. If my dog sleeping was a sign of rain, it would be raining cats and — well, he's a good dog.

Frogs

A lot of frogs have spent a lot of time in jars doing duty as "poor man's barometers" over the years. It was a long tradition in Europe, this idea that a green tree frog was especially sensitive to changes in air pressure and its behavior could predict changes in the weather. The jar was half-filled with water, and there was a little ladder. On bad-weather days, the frog stayed in the water, the thinking went, and when it was getting better, the frog climbed up the ladder out of the water.

This was just a myth, but it makes you wonder about frogs and toads and other amphibians in recent years. Their populations seem to be disappearing around the world. Are the frogs trying to tell us something?

Ants

For thousands of years, people have been looking at the industrious ants for signs that the colonies were able to foretell the weather. It seems like the Greeks had the idea first. (Then again, it seems like the Greeks had a lot of ideas first.)

"It is a sign of rain if ants in a hollow place carry their eggs up from the ant-hill to the high ground, and a sign of fair weather if they carry them down," wrote Theophrastus, a pupil of Aristotle's back in 350 B.C. The idea is that because they sense a storm on the way, the colony sets to work closing up the entrances and reinforcing their hills with new soil.

An American put the business to rhyme with the saying, "When ants their walls do frequent build, rain will from the clouds be spilled." Ants aren't the only insects busy at a change in the weather, it seems. Other old signs of coming rain: Gnats bite, crickets are lively, spiders come out of their nests and flies gather in houses.

Birds

Birds seem to have been relied on frequently as forecasters, as close as they are to the weather and all.

All kinds of weather signs related to birds of one feather or another have been passed down through the ages, although many of these are nothing to crow about.

There is an ancient Chinese proverb: "The call of the cuckoo heralds spring planting." There is old American folklore about the singing of birds. When they stop singing, for example, rain and thunder could be on the way, and if they sing in the rain, it's a sign that fair weather is coming.

In the barnyard, when chickens crow before sundown, it is a sign of rain the next day. If a rooster crows on the ground, rain is coming. If he crows on the fence, expect fair weather.

And the behavior of wild birds has been closely watched for signs of weather change at least since the time of the ancient Greeks. They thought it was a sign of rain when gulls or ducks plunged under water, and when they flapped their wings it was a sign of wind. And a heron screaming was a sign of wind. And the ancients watched for nesting activity as a sign of the seasons to come. If they flock together in moderate numbers on an island, it is a sign of good weather, but if the flock is big, drought could be on the way.

Caterpillars

The Wooly Bear Caterpillar is a standard of American weather folklore for some reason. The idea is that this caterpillar of the tiger moth is able to foretell how severe the coming winter will be.

Every autumn, the dense coat of fine hairs of the wooly bear gets a lot of press attention. The width of the brown band around its middle is carefully measured as an indication of what to expect of winter. And every spring, people look back at the predictions divined from the caterpillar's wooly coat, and it's almost never verified — a complete fantasy.

Squirrels

The scurrying little critters in the forests and the trees have done pretty well for themselves as subjects of seasonal weather forecasts.

Chipmunks were watched around the Great Lakes, for example. If they were tucked away for the winter by October, a cold and long one was on the way. If they were seen in the forest until December 1, winter would be short and mild.

And squirrels were popular signs of the coming winter. If they were seen laying away an especially large food supply, a long and severe winter could be expected. This widely held idea inspired a rhyme: "When he eats them on the tree, weather as warm as warm can be."

Some killjoy did a 20-year study of the squirrels back in the 1880s and splashed cold water on the whole idea.

Groundhog

Who hasn't heard the old one about Groundhog Day? This is evidently an ancient German tradition, but judging from all of the media attention every February 2, Americans just love the idea.

If the hibernating groundhog pokes his head out of his hole on that day and sees his shadow, he heads back into his burrow knowing that he faces another six weeks of winter. If he doesn't see his shadow, winter will be over soon.

This date marks the halfway point between winter and spring, a time some years when just about anything good you can say about the weather is welcome.

Livestock

You've got to give those shepherds their due. Anybody who spends that much time looking at all of that livestock deserves to be listened to, and the tenders of the flocks have been considered experts on weather changes for centuries.

If sheep climb the hills and scatter, it means they're expecting fair weather, and if they bleat and seek shelter, snow is on the way. Cows are the same way. If they refuse to go to pasture in the morning, it will rain before nightfall. If a cow stops and shakes her foot, there is bad weather coming behind her.

Watch the way they lick and scratch. It means rain is coming if cattle lick their forefeet, or lie on the right side, or scratch themselves more than usual against posts or other objects.

For some things, you just have to take the word of the shepherds.

Fish

Maybe you've noticed this: People who fish a lot have a certain amount of spare time on their hands. They certainly have time to think about the weather and to look for signs of change in the fish.

According to one tradition, fish bite readily and swim near the surface when rain is expected. But according to another, fish are inactive and won't bite just before thunder showers. Some not only know bad weather is coming, but where it is coming from. The saying is that blue fish, pike, and others jump with their head "toward the point where a storm is frowning."

Chapter 18

Ten Grand Old Weather Proverbs

• •

In This Chapter

▶ Taking a proverbial look at ten old favorites

• •

The weather proverb business has fallen on hard times. There's a lot of the old stuff around to ponder over and puzzle through, but darned if you can find a new one anywhere. Weather science has done that to the proverb trade. You don't hear your favorite forecaster going on at length these days about red skies at night or mare's tails and mackerel scales and what-have-you.

It's just as well that the forecasts you hear are grounded in more solid stuff nowadays. Science has taken a lot of the surprise out of the weather, of course, but along the way it's taken some of the charm.

In this chapter, you can take a look at some of the oldest and best of the weather proverbs that have been passed down. For many centuries before the science of meteorology came along, generations of weather watchers shared these sayings that sailors and farmers and other close observers of the sky found true, or at least helpful, in figuring out what was going to happen next.

Think of it as a little Weather Proverbs Hall of Fame. And you'll get an idea of their weather sense and see why some of them have stood the test of time.

Red Sky at Night, Sailor's Delight . . .

Red sky at night, sailor's delight.

Red sky in the morning, sailor's warning.

This is one of the oldest and one of the best. The ancient Greeks had a variation of this saying. And in the Bible, Matthew 16.2-3, Jesus says to the fishermen, "When it is evening, you say, 'It will be fair weather, for the sky is red,' and in the morning 'It will be stormy today, for the sky is red and threatening.'"

Here's the weather behind it: If the sky is red when the Sun is setting in the western sky, often it means that the sunlight is passing through dust particles in a clear sky. And because most weather travels from west to east in the middle latitudes where most people live, chances are good that the next day the sky overhead will be clear. On the other hand, when the eastern sky is red at sunrise, it means that the high pressure and fair weather has passed, and low pressure, clouds, and storms could be threatening.

Clear Moon, Frost Soon

This old saying picks up on a common occurrence, but isn't exactly foolproof. If the sky is clear enough to make the image of the moon especially clear, the surface of the ground will rapidly radiate away the day's heat and its temperature will fall dramatically. If temperatures get cold enough and the air is stable, frost can form on a clear night. If temperatures don't fall to freezing overnight, however, and wind kicks up, well, don't count on this one.

Early Thunder, Early Spring

Thunderstorms require a certain amount of heat. The air needs to be moist and warm, as a rule, for clouds to generate the kind of growth and commotion for lightning and thunder to result. If it comes in the winter, earlier than usual, it may mean that the air contains more heat than is common for many winter storms, and so spring will be coming sooner rather than later.

But this proverb points out one of the weaknesses of this kind of weather forecasting. Many of them are true to one location, but not another. They don't travel well. In many regions, thunder can be heard just about any time of year.

After Frost, Warm . . .

After frost, warm.

After snow, cold.

This is an old Chinese proverb that works pretty well, but not perfectly. Frost often appears under clear skies and high pressure after cold air has moved through and there is a reasonable chance that the following day will be clear and warmer. Snowfall, on the other hand, often is followed by a day that remains cold as the air mass that produced it lingers over the area. This is another saying that you wouldn't want to take to the bank.

Mare's Tails and Mackerel Scales . . .

Mare's tails and mackerel scales make tall ships take in their sails

Mare's tails are those wispy high cirrus clouds that often are followed by a thickening layer that forms a bright clumpy pattern that might remind you of the scales of a fish — a mackerel, say. These cloud formations often come in advance of thicker and lower clouds that mark the arrival of a warm front. The appearance of these clouds can be followed by veering winds that sailors of tall ships would have to contend with, and eventually by rain.

Rainbow in the Morning . . .

Rainbow in the morning gives you fair warning.

The effect of a rainbow depends on sunlight reflecting off of falling raindrops and bouncing back toward your eyes. The Sun needs to be at your back. If you see a rainbow in the morning, when the Sun is in the east, it means that the rain is falling in the west. Because weather generally travels from west to east in the middle latitudes, a rainbow in the morning means that rain is coming your way.

When Halo Rings the Moon . . .

When halo rings the moon or sun

Rain's approaching on the run.

The halo that forms around the moon and the sun is the result of its light shining through a veil of high cirrus ice-crystal clouds. These clouds are often the sign of an approaching warm front of air. The thin cloud layer will be followed by a thickening cloud layer, and eventually by rainfall. Not every veil of cirrus clouds is followed by a thickening layer. Sometimes, a storm may be bypassing you, but you will see the fringe clouds that could cause a halo. So, don't count on this one too often.

Rain Long Foretold . . .

Rain long foretold, long last.

Short notice, soon past.

This proverb tells you something about how the big mid-latitude storms of winter are generally shaped. Rain long foretold is typically storminess brought on by the passage of a warm front. This storm brings a big cloud layer that is overhead for several hours before rain begins to fall. A cold front storm, on the other hand, is a faster-moving creature and is not so easily foreseen. Its clouds arrive quickly and its rainfall comes sooner and falls more heavily, but seldom lasts long.

A Year of Snow, a Year of Plenty

A warm, thawing January or February may cheer the hearts of city slickers who are tired of winter's weather, but for many a farmer, it's bad news. It means that grains and other crops will begin to sprout too soon, and the young growth will perish in a killing frost. "January warm, Lord have mercy!" is an old saying that captures the sentiments of the farmer. "If grain grows in January," says another, "there will be a year of great need." A year of snow, a year of plenty means to the farmer that the growing season will unfold the way it should, and the land will have plenty of nourishing moisture through the spring. This proverb has real staying power in the countryside.

In Like a Lion and Out Like a Lamb

When this proverb is applied to the month of March, as it often is, it doesn't work very well. In the middle latitudes, the idea is, March 1 is still in winter, with its roaring cold winds. But the end of March is in spring, which is supposedly gentler, like a lamb. A weather scientist will tell you that in many parts of the world, most of the time, the reverse could just as easily be said of March. Applied more generally, however, this proverb describes the old and widely held idea that the great contrasts of weather and climate — between heat and cold, calm and storminess — tend to balance themselves out over time.

Appendix

Internet Resource Directory

• •

• •

*W*eather-related Web sites, newsgroups, e-mail discussion lists, and other services are an enormously large presence on the Internet. Thousands of servers are devoted to the subject, offering everything from live pictures from *webcams,* digital video cameras perched outside of television stations and other buildings around the world, satellite and radar images, output of numerical forecast models, and raw climate and weather data intended for professionals.

This list of Internet resources is by no stretch of the imagination intended to be comprehensive. Instead, the list contains a relative handful of Web sites that can serve as major portals to other material throughout cyberspace. Good luck!

Government Web Sites

The Internet has become an important communications tool for real-time weather and climate information by federal agencies. That means you and I get access to official weather data and information that used to be transmitted across the country in ways we could see. It's all there to look at now: the real-time warnings of severe weather events around the nation, the forecasts, the output of numerical models, climate predictions — the works. For a weather enthusiast, it's a feast!

National Weather Service

`weather.gov`

A good, noncommercial place to begin is Weather.Gov, the home page of the National Weather Service's Interactive Weather Information Network. This Web site is a launch pad to some of the coolest weather stuff on the Internet, including dazzling satellite images and other weather graphics as well as National Weather Service warnings of weather hazards around the United States that are updated every five minutes.

This site includes a link to the National Weather Service home pages, which open the cyberspace door to the regional U.S. weather servers as well as links to all the NWS Aviation Weather Center, Climate Diagnostics Center, Climate Prediction Center, National Hurricane Center, National Severe Storms Laboratory, and the Storm Prediction Center.

National Oceanic and Atmospheric Administration

`www.noaa.gov`

Another especially interesting government Web site is the home page of the National Oceanic and Atmospheric Administration, the parent agency of the National Weather Service, which offers a wide range of climate and weather information.

The NOAA Web site also offers links to authoritative discussions providing background and official updates on such topics as tornadoes, hurricanes, fire weather, El Niño, and drought. The NOAA weather page also gives easy access to satellite and radar images and forecasts.

University Web Sites

Good Web sites at research universities around the nation and the world are so numerous it is almost unfair to single out any. There is some very interesting weather research underway, and universities continue to expand their internet sites to describe this work. Consider this a mere sampling of what's out there, and check them out for links to other great Web sites.

University Corporation for Atmospheric Research

```
www.ucar.edu
```

UCAR is a nonprofit corporation formed in 1959 by research institutions with doctoral programs in the atmospheric and related sciences. UCAR was formed to enhance the computing and observational capabilities of the universities, and to focus on scientific problems that are beyond the scale of a single university.

At the UCAR site, check out the National Center for Atmospheric Research and the real-time weather page provided by the Research Applications Program:

```
www.rap.ucar.edu/weather/
```

This site provides ready access to some of the computing and observation products available at the research center as well as an interesting page of links to other weather sites.

University of Michigan Weather

```
cirrus.sprl.umich.edu
```

UM Weather, as this Web site is called, has the Net's largest collection of weather links. UM Weather's famous WeatherSites page is worth stopping by to check out. This page provides access to over 380 North American weather sites.

Pennsylvania State University

```
weather.psu.edu/weather/home.html
```

Penn State has a famous weather Web site that provides data for some serious weather watching. The site provides access to current upper winds observations, the output of the major numerical forecast models, as well as an interesting set of answers to frequently asked questions about meteorology.

University of Illinois, Urbana-Champaign

`www.atmos.uiuc.edu`

The Department of Atmospheric Sciences at the University of Illinois at Urbana-Champaign has developed an extensive weather Web site that includes an innovative tutorial of internet-based teaching. In addition to its wide variety of resources, check out its World Weather 2010 Project.

Special Resources

Two documents available on the Web are famous for the amount of valuable information they provide about weather. Not only to they contain a great deal of information, but they have compiled enormous lists of other resources, including Web sites and books on the subject. They are called FAQs — the internet's shorthand for answers to frequently asked questions.

Meteorology FAQ

`www.faqs.org/faqs/meteorology/faq-intro/`

This is a series of FAQ postings for the Usenet newsgroup sci.geo.meteorology. FAQ stands for Frequently Asked Questions. This particular seven-part FAQ is a remarkably comprehensive catalogue of resources available to answer questions about meteorology. The bulk of this FAQ series is about data sources, but a lot of other information has been added.

Hurricanes FAQ

`www.aoml.noaa.gov/hrd/tcfaq/tcfaqHED.html`

Research meteorologist Christopher Landsea, at NOAA's Hurricane Research Division in Miami, Florida, has developed an extensive FAQ about Hurricanes, Typhoons, and Tropical Cyclones. It contains definitions, answers for some specific questions, information about the various tropical cyclone basins, provides sites that you can access both real-time information about tropical cyclones, what is available online for historical storms, as well as good books to read and various references for tropical cyclones.

Commercial Web Sites

Cable television and broadcast networks and commercial weather forecasting companies are a big and valuable source of internet weather information. Virtually all of them offer access to local weather forecasts around the nation and the world as well as frequently updated news and features about weather events making headlines.

AccuWeather, Inc.

```
www1.accuweather.com/adcbin/index?partner=accuweather
```

AccuWeather's Web site provides comprehensive worldwide weather information, including detailed 10-day forecasts, a local ultraviolet index, and the company's own "RealFeel Temperature" index. AccuWeather offers U.S. radar images and animations and an extensive day-by-day almanac of normals and past weather information. It has many video features as well as up-to-date and helpful background information about hurricanes, severe weather, and winter weather.

Cable News Network

```
www.cnn.com/WEATHER/index.html
```

Cable News Network's Weather Main Page offers a quick look at top weather stories. Its Storm Center page provides easy-access background information and safety tips on major weather threats such as hurricanes, tornadoes, and severe thunderstorms.

Intellicast

```
www.intellicast.com/
```

Intellicast, a product of Weather Services International, is a highly developed site of current weather information tailored to travelers and other special users of weather information as well as good background. This site's Almanac feature offers helpful descriptions of month-to-month weather patterns across the nation and among the various regions of the country.

The Weather Channel

www.weather.com

The Weather Channel maintains a large and hugely popular site for up-to-date weather news, current national, regional and local forecasts as well as an extensive and well-organized collection of helpful background information. Its section How Weather Works is a helpful look at the basics of weather and its Storm Encyclopedia is an extensive presentation of important weather and climate topics. A Weather Glossary defines over 800 weather terms. The site also includes streaming video clips of Weather Channel televised segments.

USA Today

www.usatoday.com/weather/wfront.htm

USA Today has a well-developed site that presents current weather information and forecasts. It's on-line Weather Almanac feature offers concise compilations of climate data for hundreds of U.S. cities as well as an extensive collection of well-presented historical weather information. It is a highly informative and comprehensive site.

Newsgroups

If you're willing to put up with the off-subject "noise" and the sometimes rude "flaming" arguments that occupy so much of the Internet's Usenet newsgroups, a lot of helpful information and news of current weather events is available. Others may prefer more organized e-mail discussion lists devoted to weather.

Sci.geo.meteorology is probably the most active newsgroup among several newsgroups devoted to discussions of weather events and issues. Others include

 Alt.talk.weather

 Bit.listserv-wx-chase

 Bit.listserv.wx-talk

 Clari.news.weather

 Ne.weather

 NCAR.weather

E-mail List

E-mail discussion lists are an easy and interesting way to share your interest in weather with other enthusiasts and experts around the world. Here's how to get started with two leading e-mail weather lists:

✔ WX-TALK, an active, broadly based e-mail discussion list devoted to weather, is maintained by the University of Illinois at Urbana-Champaign. For information about the e-mail discussion list WX-TALK and other specialized weather-related lists, e-mail to `listserv@po.uiuc.edu` the following message: sendme wx-talk.doc.

✔ CASILIST, is an active e-mail list maintained by Jesse Ferrell for the Central Atlantic Storm Investigators organization. For more about this group, and how to join the CASILIST, check out the Web site:

`www.weatherwatchers.org/members/casilist.html`

One Final Site . . .

One of the photographers for this book, Jim Reed, has a cool Web site you'll want to check out as well:

`www.jimreedphoto.com`

With American Photo, CNN, the New York Times, Popular Science, and the World Meteorological Organization among his credits, Jim Reed is recognized as one of the finest severe weather photographers working today. A professional storm chaser, Jim has spent the past decade documenting more than 100 record-setting storms including blizzards, floods, tornadoes, and the direct strike of four major hurricanes. Enjoy!

Index

• *B* •

• R •

FOR DUMMIES
BOOK REGISTRATION

Register This Book and Win!

We want to hear from you!

Visit **dummies.com** to register this book and tell us how you liked it!

✔ Get entered in our monthly prize giveaway.

✔ Give us feedback about this book — tell us what you like best, what you like least, or maybe what you'd like to ask the author and us to change!

✔ Let us know any other *For Dummies* topics that interest you.

Your feedback helps us determine what books to publish, tells us what coverage to add as we revise our books, and lets us know whether we're meeting your needs as a *For Dummies* reader. You're our most valuable resource, and what you have to say is important to us!

Not on the Web yet? It's easy to get started with *Dummies 101®: The Internet For Windows® 98* or *The Internet For Dummies®* at local retailers everywhere.

Or let us know what you think by sending us a letter at the following address:

BESTSELLING
BOOK SERIES

For Dummies Book Registration
Dummies Press
10475 Crosspoint Blvd.
Indianapolis, IN 46256